The Essential Guide to Bulbs

the ESSENTIAL GUIDE to BULBS

Grow a Bounty of Beautiful Bulbs in Gardens and Containers

Jenny Rose Carey

Timber Press | Portland, Oregon

I dedicate this book to all the people who have made the garden here at Northview, where many of the photographs in this book were taken. To Wilmer and Anna Atkinson, who built the house in 1887, began the garden, and planted amazing trees. To Joe Giampa Jr., and Hanna von Schlegell, who care for the garden today. And to my family, who have helped me plant and have added their own touches of creativity to the growing garden.

CONTENTS

Feelings about certain bulbs differ from gardener to gardener and change over the decades. Dahlias are one bulb flower that divides opinions. This collarette dahlia is called 'E Z Duzzit'.

INTRODUCTION

The Brilliance of Bulbs

Trends in gardening come and go. By the time you are reading this, I have no way of knowing what will be fashionable in gardens, but I do know that people will be growing some of the bulbs in this book. Gardeners have loved growing flowers from bulbs for centuries; it's a trend that has always been in fashion, so to speak. During that time, certain bulbs have been "in" during one decade, and at another time, "out." How about big blowsy dahlias in vibrant colors? Or tiny snowdrops? What do people think about those *now*, at the time you're reading this? Whatever the specifics of the current obsessions, there will be gardeners who grow bulbs. I know this because bulbs are such good garden plants that we just *must* have them.

There are several features that give bulbs such universal appeal. The first is that they are extremely diverse and there really is something for everyone. You can choose from an extensive variety of colors or shapes—there are tall ones and short ones, and ones for every season. They also have the wow factor that every gardener wants. When bulbs are in bloom, you get so excited that you invite your neighbor round for a cup of tea and say, "you have to come and see my lilies!" The best part of all is that bulbs are easy to grow. If you are a beginning gardener, you can plant bulbs. And if you are an experienced gardener, you will still want to plant bulbs because there are so many new ones to try.

This book is borne of a lifetime passion for bulbs. I have always thought that bulb planting was somewhat miraculous. First when watching my older family members, and then when I was able to help them, it seemed amazing to be able to pop little brown objects into the ground that sometime later became a flower. What a crazy idea that was to my young brain! I remember a love of bulbs that goes back to the patch of purple rock garden irises outside my granny's door, the feeling of running through bluebell woods, and the scent of a freshly picked daffodil from long ago. Once I was lucky enough to have my own garden, either rented or owned, my gardens really began. I have grown many things over the course of my gardening life, but I always come back to bulbs.

We want our gardens to look beautiful, be full of flowers, and be humming with pollinators—bulbs contribute toward all these aspects. We also want to have flowers in our gardens for many months. Bulbs can help with that too. When planted correctly, bulbs bloom in sequence, one after another.

The bulbs described in this book are easy to grow in most climates. Not all of them will grow in every garden because of different soil types, climate, or weather. However, there are so many bulbs to choose from that, with some research and trial and error, you will find the ones that do well in your garden.

You will also find that your tastes evolve and change with the years. I find that I have some favorite bulbs that I grow every year, while others are with me for a few years and then I move on. All inhabitants of your garden do not have to stay with you forever. My garden is my canvas. It is a painting that is never finished, much to the incredulity of my very patient husband. I play in it. New plants are added, others are taken away. I will never be satisfied with my garden, but it is fun trying to make it more to my taste. Most of my pleasure is in the doing and not in the standing back and saying, "There—look, I am done …"

Some consideration and planning are necessary to have bulbs that fill your garden or pots with color and beauty for several months of the year. It also requires that dreaded thing called "patience," as more than anything, planting bulbs into the dark soil is an act of faith, the hope for a future garden that is bright and colorful. Weeks or months may pass before you see the fruits of your labors. I am not a patient person in many aspects of my life, but somehow gardening with bulbs has taught me a modicum of patience combined with the joy of anticipation. In other words, bulb growing is the opposite of instant gratification. Unless you go and buy a pot of daffodils in full bloom or a luscious lily that is ready to be put straight into your garden, bulbs do not produce an instant effect. You plant them, they take their time to appear, and then they give you the show that you have been imagining for months.

To get a full and floriferous garden, every square foot of garden space, or each container, must be used several times over during the year. Each set of bulbs comes up in turn and is then followed by those that bloom later. Maybe the bulbs coexist and live in the ground together, or maybe you add and subtract some bulbs. Whatever methods you use, you are already planning what comes

Grow what you love—I love these 'Apricot Parrot' tulips.

next, even as you are enjoying the current bulb display. This reminds me of one of my favorite expressions that says, "the seasons roll." This is so true when describing how bulbs grow. They each have their time of flower and their time of rest. If they are hardy, expect to see them at the same time next year. They are markers of the season; we look forward to seeing them and thoroughly enjoy them when they are out.

As the gardening year gets underway, I find myself desperate for some color and fragrance. I could not think of my late winter and early spring garden without my snowdrops and crocuses. These are the brave bulbs, if I were to personify them. They are the survivors of the plant world that bloom while nasty cold winds bring in snow flurries.

Of course, I am then looking forward to the first daffodil, the earliest tulip, and the scent of hyacinths. Alongside these indicators of the early spring moment are the lesser-known little bulbs that are not less important, but just taken for granted: muscaris, scillas, ipheions, and more. These are followed by the late spring show of large tulips and alliums. The spring bulb spectacle is when your color daydreams can become reality. The enjoyment of color is a fleeting but extraordinarily beautiful facet of gardening. When I go out to my garden on a sunny day, and there in front of me is a patch of bright-orange and pink-flamed tulips, paired with a purply blue hyacinth, it gives me a great mental lift and makes me want to sing a happy song.

The next big burst of bulbs comes in summer. By then, the lilies are in full swing. Lilies alone can be the star of your garden from early to late summer. If you have not tried them, have a go. Alongside these flowers are some amazing large leaves that burst into prominence at this time of year and add so much to the summer picture. Leafy colocasias, alocasias, and caladiums grow from tropical and subtropical bulbs. For more variety of color and shape, add in cannas, begonias, and gladioli. Just a few of these vibrant plants will add a touch of glamour to your summer garden. Dahlias come along later in the summer. They are the queens of this season with their generosity and diversity. To me, these late flowers are some of the most sumptuous of the year.

The bulb show is not done yet. The final hurrah brings fall crocuses, colchicums, and sternbergias that are new to the scene. The bulbs of summer continue their show alongside these newcomers and keep going until frost.

This snapshot of the bulbs of the year is not an exhaustive list, but it gives an impression of the cyclical nature of bulb growing. As you learn more about bulbs that are good growers in your area, you will get an idea of the best ones for your garden. You will make up your own bulb cycle. You will learn how to keep the show going and when to plant the bulbs to achieve that display.

I always get excited about growing bulbs. Yes, I do hear my husband in my ear saying, "more daffodils?" as I click "place order." But I am still fascinated by them after all these years. I think that I find bulbs enthralling because you do not exactly know what you are going to get out of the bulb. It is the thrill of the big reveal when the bulb comes into bloom. Whatever the reasons for growing your own bulbs, I want to encourage you to discover a few brilliant new bulbs to grow.

A Note on the Structure of This Book

Bulbs are fantastic and easy plants to add to any garden. But this book is set up so that you can create your own pathway to learning about bulbs. The chapters have themes, but they can be read in any order. If you are new to bulb growing or need inspiration, I recommend starting at the beginning. In chapter 1 you will find out why I am so passionate about incorporating bulbs into all my garden areas, in addition to bulb basics, from history to types and definitions. Chapter 2 contains detailed plant descriptions and photographs of bulb species and cultivars. This is divided into two seasonal parts, arranged alphabetically by flowering time. The first part shows autumn-planted bulbs that bloom from late winter through spring, and the second part features spring-planted bulbs that bloom from summer through autumn. Choosing the bulbs for your garden first will help you as you plan.

But, because each garden and gardener are different, chapter 3 is a resource for personalizing your bulb garden with design ideas and plenty of photographs. And because bulbs are perfect for growing in containers, chapter 4 is wholly devoted to this subject—read this if you want to add some potted bulbs to your garden or if you only garden in containers. Chapter 5 includes the nuts and bolts of buying and growing healthy bulbs. If you want to know how to grow bulbs well, that's where you'll find more in-depth advice.

No matter how you read this book, I hope that it will bring you pleasure and encourage your bulb-planting adventures.

I love this wonderful small bright pink collarette dahlia called 'Sean C'.

1

THE JOYS OF BULB GROWING

grow bulbs because they bring me joy. There is nothing that can beat the sheer pizzazz that bulbs can bring to your garden. If you have a lackluster corner that needs a refresh, think bulbs. Either in the ground or in a container, a seasonal planting of bulbs can cheer you up in no time. I plant bulbs everywhere. There are bulbs for sunny places and shady ones. There are some that thrive in wet or dry soils. Each bulb has its own flowering season, so you can find a bulb or two for every time of the growing year.

This is a personal look at bulbs that I love, but also ones that I think you might love too. Throughout the book, you may be able to read between the lines and tell which ones make my heart sing. I find other bulbs useful, but less emotionally appealing. I have my favorites as every gardener does, but I am a big proponent for making your garden match you as a person. I might express what I like, but I do not want you to like the same things. In fact, I would be very

disappointed, in a few years' time, if I come to visit your garden and it looked the same as mine.

The main criterion for being included in this book is that the plants are usually bought as bulbs. There are cases where they can be purchased either as potted plants or as bulbs, but even if they are in containers, there is a bulb in the soil. These bulbs are easy to grow in a garden or in a container and are widely available. Remember, of course, that not all bulbs are equally simple

1. This river lily is not a lily, it is a *Hesperantha*. 2. This red spider lily is also not a lily, it is a *Lycoris*, here under-planted with a red-and-white caladium. 3. Another non-lily that is called pineapple lily, which is *Eucomis*. 4. Finally a real lily that is in the genus *Lilium*. This lovely one is an Asiatic 'Yellow Cocotte'.

to grow in every garden as your climate, weather patterns, topography, and soils are unique.

How to Use This Book

This book is a guide to bulb growing that is based upon my real experience of many years of gardening. I have gardened in two places: England and the East Coast of the United States. I also travel to visit gardens in other places, taking notes and photographs as I go. Giving lectures provides another window into gardening lives, in different climates and situations.

As you read this book, imagine that we are walking around my garden here at Northview. It is a lovely day in southeastern Pennsylvania, and we are chatting about bulbs that we want to grow. I wrote this book as a way of sharing my thoughts with you about how to choose, plant, and care for your bulbs.

I have used the same commonsense approach when thinking about how to name the bulbs in this book. Each bulb has been called different names by different people over the course of time. Gardeners and botanists have tried to standardize plant names so that we can communicate with people in other countries and know that we are talking about the same thing. In the interest of the same clarity, I default to the use of Latin names in the plant palette, so that we can be sure that we are talking about the same plant. However, I do not mean to downplay other names that these bulbs have been called during their history with humans. I love the so-called common names that often have a musical or romantic sound to them and are meaningful to the people who use them. Unfortunately, we humans are not very inventive, as one gardener's bluebell is not necessarily the same as another person's bluebell. And when you begin talking about a lily, you could mean one of many dissimilar plants. I have included common names in the plant descriptions, as well as the Latin ones.

An Ecological Ethos

I have gardened all my life. During those many years, I have realized that when I garden, I want to do so in harmony with the flora and fauna in my garden. I still get mad at the squirrels for digging up my bulbs and I let them know it. But apart from these outbursts, I endeavor to take a big-picture approach. My garden is one little link in a chain of green spaces that we call gardens; these small plots link together over back fences and side yards, through fields and woods, to create green corridors across the world. They are also the part of the earth where we can make our own choices for a positive future.

I choose to stay away from any chemicals that may harm the modest ecosystem in my garden. I have created an environment for my family, pets, and visitors that is as safe as I can make it. I do not spray insects, so that animals that are insectivores have things to eat. The herbivores have a lovely selection of choices too. There are so many edible plants here that my husband refers to my garden as the all-you-can-eat salad buffet. I want the birds, frogs, toads, skunks, rabbits, deer, turtles, snakes, raccoons, insects, and other unnamed creatures to have healthy things to eat. Maybe just not my most precious bulbs, please.

OVERLEAF: This spring garden at Northview is packed with daffodils, hyacinths, and muscaris.

A WORD ON NAMING

Each plant and animal in the world has a unique double name in Latin to provide consistency across languages—if we all use the Latin, then supposedly we're all on the same page. The Latin name is italicized to show that it is in another language. The name sometimes also includes a third word, to give even further specificity.

The first word of the Latin name is called the genus. This is similar to your last name and starts with a capital letter. All plants within a genus share similar characteristics. With bulbs, this name is often the same or very similar to the English name. For example, *Crocus* in Latin is also crocus in English. Dahlia is also the same in both languages. Even tulips are in the genus *Tulipa*, which is easy to guess.

The second name is the species name and has a lowercase first letter. This is used to tell the members of the genus apart. The Latin second name compares to your first name and often (but not always) describes some aspect of the plant. For example, within the large genus of *Fritillaria*, one tall species is *Fritillaria imperialis*, the crown imperial fritillary. A smaller species is *Fritillaria meleagris*. The word *meleagris* means that it is spotted like a guinea fowl. If you look at this flower, it has patterns that resemble the feathers of these birds.

The final name shows the cultivar or cultivated variety. This is written in the local language, surrounded by single quotation marks, and is not italicized. A lot of bulbs are known just by the genus and a cultivar. They have been bred from several different species and their heritage is not fully known. For example, *Tulipa* 'Ballerina' is the full name of the orange ballerina tulip.

Some cultivars are produced by chance as a naturally occurring mutation, or sport, among a field of flowers. In other cases, bees or other pollinators crossed one plant with another to produce something unique and worth propagating. There are also bulb breeders that are producing new cultivars by deliberately crossing one desirable bulb with another closely related one.

Not all bulb names have the third cultivar name. Species bulbs will just have the two Latin names. This shows that they are the same plants you could find growing in their natural habitats. But that does not mean that they are not good garden plants—many of them are. They are great to use in wildlife gardens, especially if the bulbs are native to your growing area.

Another positive choice that I made a long time ago was to actively garden for pollinators. It is an important fact, if we as gardeners do not include lots of flowers in our gardens for pollinators, they will continue their worrying decline in numbers. The good news is that many bulbs have flowers that are perfect for pollinators. In my garden, I have so many types of pollinators that I cannot identify them all or count them. Many of the photographs for this book were taken here at Northview, and as we looked at each photo closely, there were insects large and small in many of them. You may notice them too.

Growing Bulbs for Pollinators

As you choose your bulbs, think of adding some that will encourage pollinators to use your garden. Include a variety of flower shapes and colors that will be enticing to a wide range of pollinating animals. Choose native bulbs, when possible, as they will invite beneficial insects, too. A mass of flowers together is highly visible to pollinators as they fly by. Repeat bulbs and plant them in patches or swaths. Once pollinators start feeding, they can efficiently move from flower to flower without wasting too much time or energy. The appearance of plants grouped by type is attractive to us *and* to passing pollinators.

Attempt to have flowers in bloom for as long as possible during the year to serve any pollinators that are on the wing. Add fragrant flowers because the scents attract pollinators, especially in winter and in the evening when there are few other plants in bloom. There are bulb flowers to suit pollinators that bloom in sequence from late winter through fall. Some of the best bulbs for pollinators include many of the spring bloomers

TOP: Butterflies pollinate many bulb flowers as here on this red lily.

BOTTOM: This open-faced dahlia called 'Appleblossom' is a great bee attractant.

OPPOSITE TOP: This crown imperial fritillary lives up to its royal name by sporting a crown-like topknot of orange flowers. It differs vastly from other garden fritillaries in size, color, and shape.

OPPOSITE BOTTOM: The whole name of this little species tulip is *Tulipa heweri*. It originates from parts of northern Afghanistan.

like tulips, daffodils, and small bulb flowers. Later in the gardening year, pollinator-friendly bulbs include gladioli, cannas, crocosmias, irises, tuberoses, single dahlias, and fall-blooming crocuses.

Growing Bulbs That Suit Your Garden Conditions

One other positive thing that I do is match the bulbs that I choose to my garden's growing conditions. This is a basic principle of gardening that requires the least possible effort or intervention on your part. Hardy bulbs that grow vigorously are likely to perennialize. A strong growth rate provides enough energy for them to flower, set seed, and still have extra energy left to send back to the bulb. The bulb stores this energy during the resting period. In the following growing season, energy is mobilized to power next year's growth and blooms.

As you garden with bulbs, you will find that some bulbs do really well for you and settle down and multiply. These are the bulbs that suit your soil, climate, and maintenance routine. For a full garden, add more of these bulbs or ones that are similar.

Bulbs Around the World

If you took a trip around the world, you would find bulbs in many different countries and in a variety of dissimilar habitats. Nestled in stony soils up in high mountains, you would find bulbs that are covered by snow for much of the year. In deciduous woodlands, you could see bulbs growing that require some shade and organic-rich soils. In seasonally damp grasslands and meadows, other bulbs would be flowering. In tropical zones, you could see a completely different set of bulbs that are growing in year-round heat with plentiful moisture. Over the centuries, some of these bulbs have been brought into our gardens, where we try to find places that are enough like these wild habitats that the bulbs will continue to grow well.

Early gardeners used plants in their gardens that were harvested from wild habitats in large amounts. This caused the destruction of beautiful natural areas. Growing wild-collected bulbs is no longer necessary or condoned. There are many reputable sources that grow bulbs in nurseries and fields. Check the origins of your bulbs before you buy them. The label should say "grown from cultivated stock." We can choose from thousands of different bulbs that are ethically grown and do not destroy wild ecosystems.

These wild tulips (*Tulipa sylvestris*) are native to a large area of southern Europe and northern Asia, where they are found growing in grassy areas and lightly shaded woodlands.

A BRIEF HISTORY OF BULB-MANIA

Bulbs and humans have had a long joint history. The most favored early grown bulbs were those that could be dug up for food. At some point people realized that certain bulbs produced beautiful flowers. They chose the ones that they liked, dug them up, and moved them closer to their houses, so that they could admire them. The bulbs that grew best and were the most desired were propagated and passed along to friends. Later, plant breeders crossed one special flower with another to produce the extensive range of garden bulbs that we have access to today.

There have been times in history when people have been so taken with a particular group of bulbs that they were crazy about them. Tulips seem to inspire a particular passion. In Turkey, there was a long-lived obsession with slender tulips with pointed tips that began in the early 1500s. In the 1600s, the Dutch had their own mania for tulips, especially the ones with fantastic flame patterns. The speculation on tulips got completely out of hand and resulted in a tulip crash. Hyacinths had their moment when they were the "in" thing. Daffodils, bearded irises, lilies, and dahlias are still popular bulb groups today, with their own special interest societies. Bulb mania is easy to catch. Once you start using bulbs in your garden, it may be difficult to stop. You grow one or two little beauties, and before you know it, there are bulbs everywhere. Beware, this book comes with a lighthearted warning that you too might catch bulb mania.

ABOVE: A view of some historic hyacinth cultivars that have been preserved at the Hortus Bulborum, a bulb garden in the Netherlands.

OPPOSITE: You will be truly inspired to grow more bulbs if you visit public gardens that have beautiful displays. Here, a serene early morning view of daffodils and tulips at Keukenhof, a large bulb garden near Amsterdam, Netherlands.

Getting Inspiration for Your Bulb Garden

One way that gardeners get hooked on bulb growing is by seeing magnificent displays of massed bulbs planted in public gardens. Jump at any chance you get to see large bulb displays. Go to some spring displays, and then some summer ones too. This will help you understand how bulb plantings change throughout the year. You are not going to repeat the same displays in your garden. What you are looking for is inspiration. Being at the garden is not a time for practicalities. It is a time to revel in the visual beauty, the fragrances, and the pleasures of a bulb garden.

While you are at the display garden, try to find a few specific bulbs that you like the best, and maybe a couple of combinations that work well. I have always loved growing bulbs, but after going to the Dutch bulb gardens in spring, my garden exploded with bulbs for all seasons. You may find that you have a similar experience. I take lots of photos while I am there, and I also take pictures of the plant labels, so that I remember the bulb names.

As you process the experiences that you had during your garden visits, think about what was most meaningful for you. You may have been taken by the joyful exuberance of the plantings, or their sheer magnificence. There may be new bulbs that you have never seen before that you just have to grow. You may have loved a certain novel shape, some fantastic color combinations, or have been bowled over by the diversity of bulbs. The list of possible take-home thoughts is endless.

The Basics of Bulbs

The word "bulb" is used by gardeners as a general term, which includes all underground storage structures that are planted in the soil and grow up to produce a plant. Their actual internal organization may vary, but it is a useful word to describe them all.

When you look at a bulb, just like the nondescript outer coat of a seed, it gives you no clues as to what is contained within. The outer layers of bulbs are relatively plain. They may be brown, cream, or occasionally tinged with green, purple, or yellow. The surface could be flaky, shiny, or dull. As you observe the bulb, you wonder what is inside. To find out what type of flower it contains, you need to plant the bulb.

Growth and Rest: A Bulb's Lifecycle

To have success with bulbs, it is helpful to understand their lifecycle. You plant the bulb in a hole in the ground and cover it with soil. The next time that you see it is as a little green tip protruding through the dark earth. In the time between these two sightings the bulb has been resting for a while and then growing roots. These underground roots are the first thing to grow, and they enable the bulb to absorb water. As water goes into the bulb, it plumps up, and the plant begins to grow and develop. Roots continue to absorb water and nutrients and anchor the bulb in the soil as it grows. They can also reorient the bulb so that the roots are down, and the shoot faces up.

Many bulbs are adapted to growing in places where ideal growing periods are short. So, once the shoot emerges, the pace of growth is fast. The stem elongates, and the leaves unfurl to catch the

1. Continue to visit gardens throughout the year to get bulb-planting ideas. In this autumn garden, dahlias of all shapes and colors are growing together. **2.** The bottom of this dormant hyacinth bulb shows the shriveled roots from last year's growth and the little points where the new roots will emerge. **3.** Bulbs, like this white tulip, are preprogrammed to make flowers.

1

3

2

1. Each bulb has a specific time to produce flowers and leaves. In this scene from the gardens at Chanticleer, in Wayne, Pennsylvania, Spanish bluebell flowers are surrounded by the broad green leaves of colchicums that will not flower until autumn. **2.** In this garden bed, blue muscari with elongated leaves are interplanted with the mottled foliage of hardy cyclamen. **3.** Bulbs are lightweight, easy to ship, and quick to plant. They provide an efficient way to grow plenty of flowers in your garden. **4.** Lilies are one of the most dramatic of summer flowers and very easy to plant and care for too.

sunlight. The bulb uses the energy from the sun to produce flowers and more leaves. The flowers are pollinated and the plant sets seed, all before the weather changes. A change in the heat or water levels is often the signal to the bulb that it is time to shut down aboveground growth. This is a very important time in the bulb lifecycle to ensure its continuity for the next growing season. The bulb needs to send as much energy as it can, to renew the belowground bulb for next year. The replenished bulb becomes the powerhouse of stored energy that fuels the quick growth rate at the beginning of the next growing season.

The Seasonality of Bulbs

Bulbous plants are united by the fact that they live a seasonal way of life. Each type of bulb has a time of growth and flowering, followed by a time of rest. This resting time is referred to as dormancy. Bulbs are very much alive during the dormant phase of their life, so they still need oxygen and a certain level of moisture to keep their living processes ticking over. They wait beneath the ground during the bad times that are not suitable for growth. Some are dormant when the conditions are too hot or dry, but most bulbs hibernate during cold winter months. The bulb stays in its dormant period until certain environmental conditions change to ones that are conducive for growth. For most bulbs, the trigger is a change in temperature combined with available water in the soil.

Being underground protects bulbs from inclement weather and herbivore damage. The conditions in the soil even a few inches below the surface are less variable than the weather conditions aboveground. In summer, the soil is cooler and slightly moist, even in dry weather. In winter, the ground may freeze, but the bulbs are not subjected to the harsh freeze-thaw cycles that damage aboveground plant parts. Its position buried in the soil keeps the plant in place, even when its roots wither away in dormancy. Once the good times roll around again, up they grow, and the cycle is repeated.

Benefits of Bulbs to Gardeners

Gardeners love bulbs for many reasons. A main one is that we can embrace their cyclical lifecycle to plan for a succession of different flowers that appear in sequence throughout the gardening year. This gives us fresh things to look forward to. Bulbs bloom when the seasonal conditions are ideal, and then retire from the garden stage to wait it out until next year.

To the gardener, a bulb is an ideal structure for transporting and planting. This fact has led to the widespread movement of bulbs around the world, where they are planted in climates vastly different from where they originated. You hear tales of bulbs carried in pockets or sown into the hems of dresses. Sometimes taken legally, and I am sure, other times, illegally. Bulbs and their resulting plants are objects of desire and people have gone to great lengths to obtain them.

Bulbs are light in weight and are shipped to you during their dormant phase, so that there is no need for pots and soil. Their compact size means that they are easy to plant in containers, or gardens. Once they are planted, we also benefit from the natural tendency for bulbs to grow rapidly. If you buy good-quality bulbs and plant them in decent soil, within a few months they will be in full bloom.

Many of the bulbs in this book fall into the easy-care category. The information needed to make the whole plant is already preprogrammed within the bulb. There is a tiny flower bud in micro-mini size, and the leaves and roots are there in their infancy. All you need to do is provide somewhat hospitable surroundings and off they go.

The good news for gardeners is that bulbs are survivors. Bulbs are opportunistic plants that are fantastically adapted to seasonal shifts in weather patterns. This means that, whatever your level of gardening prowess, you can have success with them; bulbs inherently want to grow and flower even if where you plant them is not ideal.

Bulb Types

Gardeners use the word "bulb" to describe all underground storage structures that we grow in the garden. It is a good shorthand term that includes a variety of different botanical structures, each with the same function of food storage for plants during their dormant state.

If you look closely at a variety of different bulbs, it becomes obvious that they are a motley crew of varied structures with diverse shapes, colors, outer layers, shoots, and roots. In our day-to-day gardening, we really do not know or care about the details of the plant structure, but just want the thing to grow. This bulb primer explains a bit about the diverse bulb structures and explains any differences that you may need to know when you plant them or look after them.

Bulbs come in all shapes, sizes, and colors, but their outside appearance gives no indication of the beauty that lies within.

TRUE BULBS

While we are calling this whole group "bulbs," there is a subcategory that is called "true bulbs." This is one of the most common and widely grown subgroups. True bulbs are composed of layers that are actually swollen leaf bases. They are joined together at the bottom of the bulb by a tough base, called a "basal plate." When planting your bulbs, you may see tiny, shriveled roots coming from the basal plate. Plant this side downward because the new roots will grow from there. The pointed tip of a bulb is where the shoot emerges. This end needs to go upward. The future flowering plants are tucked inside bulbs in miniature, just waiting to grow.

True bulbs can be subdivided into those with a protective outer covering, and those without. Bulbs like daffodils, tulips, snowdrops, and hyacinths all have a covering. Leave this in place when planting but if parts fall off, plant the bulb as usual.

The other type of true bulb lacks an outer layer. Instead of a smooth outer coating, they have scales that are arranged in an overlapping pattern like tiles on a roof. The best example of these scaly types are lily bulbs. These bulbs tend to dry out faster than those with an outer layer and should be kept slightly moist, even in dormancy. When you receive these bulbs in the mail or from a nursery, plant them immediately.

True bulbs persist from year to year and get bigger with time. They make new bulbs by forming offsets on the sides of the basal plate. You will see this if you order good-quality daffodil bulbs that may have two or three "noses." If these are each large you can break them apart, but I usually leave them attached and plant them in the same hole.

TOP: Snowdrops grow from a true bulb.

MIDDLE: Tulips have a crisp covering that often falls off. Do not worry if this happens, but do not peel it off on purpose.

BOTTOM: Lilies are a true bulb, but they have no outer covering to protect them. Buy lily bulbs as fresh as you can for best growth and flowering. This lily has just been dug, divided, and will be quickly replanted.

CORMS

Corms have a different structure that has no layers but is solid throughout. They are, in fact, an enlarged stem base that stores food belowground. Corms are similar to true bulbs because they have a basal plate at the bottom where the roots emerge. Some corms have an outer covering and others do not.

Corms have a growth pattern where the original corm withers away and a new stem base is formed above or below the shriveled one. Some large corms produce little corms around the bottom, called "cormels" or "cormlets." These will grow leaves for a few years and then become large enough to make a flower. Examples of corms include crocuses and gladioli.

TUBERS

Tubers are another type of modified stem, with no basal plate or outer covering. Structurally, they seem a bit disorganized, with roots and shoots emerging from eyes in different places on the surface. Some tubers have a recognizable top and bottom, and with others, it is hard to tell how to plant them. When in doubt, plant them on their sides and they will right themselves.

Tubers have diverse growth habits. Caladiums grow from a tuber that shrinks and disappears with new ones being made for next year. Tuberous begonias and cyclamen have a different strategy and remain in place, getting bigger each year. They resemble a fat frisbee in shape that gets a bit thicker but noticeably wider. When you are buying tubers in person, pick out the larger tubers. Plant them with the hollow side up.

TOP: Crocuses have storage units called corms. Gardeners call all of these underground structures by the general term "bulbs."

MIDDLE: Gladioli also grow from corms.

BOTTOM: Begonias grow from an underground storage structure that is a tuber. It is slightly concave on the upper side where the shoots are emerging and rounded on the bottom.

TUBEROUS ROOTS

Tuberous roots are similar to tubers, but they are formed from root tissue, not stems. Some of their roots swell up to store food while others continue to absorb water and nutrients and anchor the plant in the ground. Dahlias are the most commonly grown tuberous root. All the growing buds are attached to the central stem, that is sometimes called a "crown." When you divide these plants, each part needs a section of the central stem, otherwise it will not grow.

RHIZOMES

Rhizomes are thickened storage stems that lay parallel to the soil surface. They send out roots and shoots directly from buds on the rhizome. Bearded irises are one of the most popular plants that are grown from rhizomes. Calla lilies and cannas are also rhizomes.

Bulb Hardiness and Microclimates

One of the factors in discovering whether a certain bulb will be hardy for you is knowing your lowest temperatures in the year. It is not the only measurement that will influence bulb hardiness, but it is a good first step when compiling your bulb list.

To determine which bulbs are likely to be hardy in your garden, you can look up your hardiness zone as designated by the United States Department of Agriculture (USDA), which can be found on the website planthardiness.ars.usda.gov. This ever-evolving map divides the United States into temperature bands, according to the lowest average temperatures. If you live in a different country, find out your average winter low and

TOP: Dahlias grow from large complex structures called tuberous roots. If you divided this tuberous root, each part would need to have at least one of the little green shoots that are visible on the central stalk to be able to sprout and grow.

BOTTOM: This bearded iris rhizome needs to be planted partially out of the soil.

correlate it with these standards. Knowing your zone is a big help when deciding which bulbs will be able to survive winter conditions outside in your garden. There are other factors to consider, such as your highest temperatures and rainfall patterns. There is no simple way to include this into distinct zones. Your best bet for knowing which bulbs will survive in your climate is to consult trusted local resources.

A common feature among gardeners is, for some reason, we often want to grow plants that will not do well in our garden. Maybe, we think, if we're talented gardeners we might be able to get them to grow. This is a rather strange pleasure that is sometimes called "cheating a zone."

The average lowest temperature that your garden experiences in a year gives you a general idea of what bulbs you can grow. To extend what you can grow, you can find areas within your garden that are slightly cooler or warmer. Even within the smallest garden, there are areas that differ in temperature that are called "microclimates." Observe your garden throughout the year and take note of hot spots and cold corners where you could grow bulbs that need specific conditions. For example, you might be able to grow borderline hardy bulbs on the sunny side of your house, right next to the heat-emitting foundation, where it will be warmer in winter than parts of your garden that are farther away.

Another factor that will determine whether bulbs will survive in your garden is how wet the bulb is when the ground freezes. If the ground is saturated and then it freezes, the bulb might be damaged. To help with bulb survival, look for areas where the soil stays relatively dry, like places that are up slopes, in raised beds, and those that are sheltered by an overhanging roof.

In all areas, adding some extra mulch will help insulate the soil.

Some bulbs need a good summer baking to flower well. For these bulbs, you need to find the parts of the garden that are bathed in hot, late-day summer sun. If you are in a cold climate, look for places where the snow melts first. That area would make a great place to have a flower bed with early spring bulbs because they will flower there before equivalent bulbs in your frosty spaces.

Hardy Versus Tender Bulbs

There are two main groups of bulbs that you can use in your garden. The first group are the hardy bulbs that will survive your coldest winter temperatures. These are usually planted in autumn, and then bloom at their preprogrammed time from late winter through summer. The second group are bulbs that are not hardy for you in winter. These are planted out into the garden in spring, after the soil temperature has warmed. The severity of your winter climate determines which bulbs you can grow year-round, and which ones you will have to dig up and save inside to preserve them against frost.

Once you know what zone or zone-equivalent you are in, you can begin compiling your list of possible bulbs. Divide your list into two halves. The first half will be the hardy bulbs that are planted in the fall, and the second are the tender ones that are not put out in the garden until after your last frost.

The bulk of your list should be bulbs that are hardy in your garden year-round. These will become the mainstay of your bulb garden with their reliable nature and abundant blooms. These hardy bulbs are the easiest ones to grow

1

2

1. Find out whether bulbs are cold hardy in your garden by looking at the USDA hardiness zones. *Hippeastrum ×johnsonii* comes from parts of tropical America and only survives outside in zone 7 and warmer. **2.** This deep pink *Amaryllis belladonna* is only hardy in zones 8 to 10 and grows well in the mild climate at the San Francisco Botanical Garden at Strybing Arboretum. Elsewhere, try this bulb in summer containers. **3.** Tender bulbs like this 'Mojito' colocasia are only hardy where winters are warm in zones 8 to 10. **4.** Bulbs such as this 'Ambergate' daffodil are very hardy and will grow in cold winters down to zone 3.

4

3

in your garden. Plant them with care and give that one-dollar bulb a million-dollar planting hole, to provide it the best possible start in life. Choose ones that are not bothered by pests or herbivores for the best chance of success. They are not quite set-it-and-forget plants but pretty close to that.

Most hardy bulbs are planted in autumn, with a few, like some lilies, that can be planted in spring. They will flower anywhere from late winter through autumn. If they survive your winters, you can consider them to be perennial. Perennial bulbs that grow well in gardens are the tried-and-true survivors that have been in cultivation for many years. These bulbs are remarkably adaptable and grow well in a wide range of regular garden conditions. Some hardy examples include daffodils, muscari, and snowdrops.

Some bulbs are short-lived perennial bulbs. These are ones that grow for a few seasons, and then either stop flowering or disappear. It is not always that they are not good bulbs for your garden. Large hybrid tulips are prime examples of this growth habit. They produce their most spectacular blooms when they are planted as new bulbs each year.

A few bulbs have a very different lifecycle. These are planted in late summer to bloom a couple of weeks afterward. If they are hardy in your area, they can then settle down and come back in future years. The strange part is that in autumn they grow flowers only and then produce their foliage later. Some grow winter leaves and others wait until spring to shoot up their foliage. Examples include colchicums and lycoris.

Tender bulbs are the stars of the warm-season garden, with bold leaves and abundant colorful flowers during the summer and autumn. In zones 1 to 7, they do not survive cold winter temperatures, so they are planted outside in the spring after the soil has warmed up. If you have cold winters, these tender bulbs will need to be stored in a frost-free place to ensure their survival for next year. If you do not want to save them, you can buy them anew each year and treat them as annual additions to the garden.

In zone 8 or warmer, these bulbs may be hardy outside year-round. If you garden where there are mild winters, they can remain outside as long as the bulbs do not get too wet or too cold. In climates with rainy mild winters, it is often best to dig up these bulbs even if they are technically hardy in your area. Some of the best examples are begonias, calla lilies, cannas, and dahlias.

Planning Your Bulb Displays

"Planning is the key to a successful bulb garden." Maybe or maybe not. This is often parroted in books and lectures, but, personally, I do a lot of winging it. You can have a lovely bulb display in your garden, simply from taking a trip to the garden center to buy a few bulbs you like, and then planting them. However, if you want to color coordinate your bulbs or purchase ones that are really suited to the growing conditions in your garden, then perhaps a little planning can help.

BULBS FOR EASY CARE

Daffodils, like this collection of historic ones at Northview, are hardy survivors in gardens.

Many of the bulbs in this book fall into an easy-care category of plants. The information needed to make the whole plant is already preprogrammed within the storage unit that we call the bulb. These bulbs produce showy foliage and flowers with distinctive shapes, textures, and colors with very little work from us.

If you love tender bulbs that are not hardy in your garden, choose ones that are easy for you to store over winter. This means that you do not have to buy them again next year, which saves you time and money. Plus, you can be sure of having your favorite plants again next summer.

First, Notice the Bulbs You Love

For some reason, bulbs divide opinions and inspire passions. It is fascinating that we are drawn to certain ones and not to others. Try walking around a garden with a group of gardeners. Everyone may *ooh* and *aah* over some plants, but for others there are vast differences of opinions. Some of the group will like the bold, and to my mind, garish bulbs. Others only like the tiny, elegant, or delicate bulb flowers that are hard to see unless you have your glasses on. I think that is why bulbs are so loved—there is something for every gardener.

It is not always possible to understand what draws us to certain plants over others, but each time you are wowed by a plant, pause for a moment, and think what it is that is fascinating you. Use your eyes, but also smell and handle the plant. Feel if there is a fascinating texture to the petals or leaves. Smell the flower at different times of day, and in various weather conditions, to see if it is fragrant at certain times. Bring a cut flower inside, where the heat of the house can sometimes bring out the scent. Identify what it is about the shape, color, size, or patterns that drew you to this flower to begin with.

Make Your Bulb Wish List

From the bulbs that you have observed, start a wish list of your favorites. I guarantee that the

1. Here, a late summer combination for shade that was a happy accident in my garden at Northview. The red center of this leafy tender caladium called 'Lemon Blush' picks up the reddish colors of the hardy begonia stems. **2.** One of my gardening friends loves this bright red crocosmia called 'Lucifer', and her hummingbirds love it too. **3.** This amazing hyacinth 'Splendid Cornelia' brings me much joy, with its fragrance and glorious color. **4.** A bulb that often tops my wish list is the bright magenta byzantine gladiolus, seen here at Colonial Williamsburg in Virginia.

list will outstrip your bank balance, so you will have to pare down your wish list into a "buying list" at a later date.

Once you have some ideas for your wish list, you need to see whether they will grow in your garden. Successful bulb gardening, like all aspects of gardening, is the result of a compromise between the plants that we love, and the ones that grow well in our climate, soils, and weather patterns. It takes trial and error to find out what bulbs thrive in the specific growing conditions in your garden. Talking to local gardeners and visiting their gardens can really help you see what types of bulbs do well in your area. If you like them, add them to your list.

Decide Which Bulbs to Grow

It can be hard to decide what bulbs to choose for your garden because of the incredible diversity of colors, shapes, heights, and plant forms that are available. It is a good idea to be selective, so that your garden has a cohesive look and does not look like you just picked up the random bag of assorted bulbs that was on sale. Variety can be good, but a concerted selection process produces a satisfying garden that is suited to you personally.

To help guide your bulb choices, think of two circles that have an interlocking space in the middle, like those old Venn diagrams from school. One circle contains the names of the plants that multiply and thrive in your garden. The other circle contains your favorite plants that you cannot garden without. The plants that fall in both circles, where they overlap, are the ones that will usually form the backbone of your bulb garden.

If there are no names in the crossover part, it indicates one of two things. Either your bulb

Camassias fall in the intersection of "love them" and "will grow in my garden." I need to plant more of them.

garden thrives and you do not love it, or it has bulbs that you love, but they do not grow well in your garden. If there is a large intersection, you are well on your way to defining the plant palette that contains bulbs that will grow, and also bring you joy. Plenty of other bulbs add sparkle and variety to your plantings, but it is these stalwart, reliable performers that become your signature plants.

Planning Strategies

If you are struggling to imagine what your flowering bulbs will look like when blooming, create a mood board. This will help you visualize your bulbs and their possible combinations. Find a photo of each bulb and cut and paste them into a document or onto a piece of paper. Once you have decided what to plant, this becomes a

As you decide on bulbs for your garden, look at what is in bloom at the same time and create combinations, such as this blue bearded iris and the fluffy purple flowers of *Thalictrum* 'Black Stockings'.

this organized, but I have done things like kept a record of how many snowdrops I have blooming each year. I know I have copies of all the orders that I have ever placed, but the exact details may be fuzzy. However, the bulbs that I really love, I always know by name.

As you use bulbs in your garden, you will get to know their growing requirements, and planning becomes easier. You will find your favorites that have to be in the garden every year and then others you want to try. If you find a tulip or two that you love, you may need to buy a few dozen more every year to reinforce an existing planting. I had a lovely gardening friend who adored the double pink-and-white tulip 'Angelique'. She planted some more of them around her roses each autumn. In this way, she was sure of a beautiful tulip display each spring.

Finding Places to Plant Bulbs

While you are thinking about what bulbs you want to plant, also consider places that you can plant them. Bulbs are often small in size, so that they are easily planted into little plant beds and containers. Your list may contain some bulbs that require sunny conditions, and others that need a certain number of hours of shade. For success with bulbs, try to match your choices to existing conditions in your garden.

Some ideas of places to plant bulbs include anywhere near your house, so that you see them regularly, alongside a fence or hedge where there is a little protection from strong winds, in the edge of a vegetable growing area, in a mixed flower bed, or in the partial shade of deciduous trees or shrubs. In any garden, you can also consider growing bulbs in containers. It is an easy way to ensure that the soil conditions you provide for them matches their growing requirements.

useful record. It helps you to remember what you planted and the names of the bulbs when they are in bloom. You can keep these sheets or online records in a garden notebook. Then, when you look at it next year as you are planning what to plant, you can decide what to repeat, and what you want to change.

Some gardeners are serious planners. They are the ones who have an organized spreadsheet to tell them every bulb that they have purchased, which bed it is in, and when it has bloomed every year. If you are a kindred spirit, and tables and data make you happy, then a bulb chart is in your future. The columns on the chart can be the times of the year when you would like to have blooms. Down the side, list the bulbs that you want to grow, and find out when they bloom. If you like to add more details, you can put colors, heights, and uses such as whether they are good pollinator plants or cut flowers. I am not quite

1. Snowdrops are the first bulbs to bloom in my garden and a sequence of them takes the garden from late autumn through spring. **2.** At the end of the growing season, this anemone-flowered dahlia blooms alongside fall asters. **3.** 'Sun Disc' daffodil is a cute, small-flowered cultivar that is one of the last to bloom in my garden. **4.** Plant your gladiolus bulbs, like those of *Gladiolus primulinus* 'Las Vegas', in batches every few weeks, so that you will have flowers all summer long. **5.** Plant little early-flowering irises like 'Rhapsody' in a sheltered, sunny corner, to get them to bloom extra early in the spring.

Extending the Season of Bloom

There are bulbs for every season of the year, so think about what season you would like to have bulbs blooming, as you assemble your lists. Some people like lots of spring bulb flowers and others really lean into summer bulbs. You may want to start with just spring bulbs until you get the hang of things; but remember that every season can be considered bulb season.

I love to have flowers in bloom in my garden at Northview for as many weeks of the year as possible. Bulbs allow me to extend my gardening year to the shoulder seasons—the quieter, more contemplative months outside of the main gardening rush. My bulb year begins in late winter with snowdrops and ends the growing season with colorful dahlias and cannas.

If one of your goals is to have bulbs of interest for many months, there are a few strategies to use. The first is to choose bulbs that will come into flower or leaf in every part of the growing season. Within each bulb type, purchase early-, mid-, and late-blooming bulbs. Take daffodils for example. The earliest daffodil to bloom in my garden is 'Rijnveld's Early Sensation' in late winter, and the last one is 'Sun Disc', some months later in late spring. This is not a bad blooming show for one genus of plants. Do the same process with crocuses and tulips, and your spring garden will be a flowery place for months. Another idea is to succession-plant. Use the same piece of ground to plant a series of bulbs that will bloom one after another. Early spring bulbs can be followed by those for later in the year.

Some bulbs are programmed to bloom a certain number of days after planting. Gladioli are a good example of this. Unless you want the whole bag of bulbs to bloom at once, try planting them in stages. Pop in a few bulbs every few weeks during the summer, and you will have some gladioli in flower for several months.

As discussed previously, you can also use the warm microclimates around your house to extend your garden's bloom. Find the sunniest spot in your garden, and plant bulbs there. What you are looking for is a little suntrap that warms up first in spring. It may be on the sunny side of your house or on a sun-drenched bank. Bulbs planted in these places will come into bloom a week or two before the ones on the shady side of that same bank. This works well with late winter and early spring bloomers, like snowdrops and rock garden irises.

These warm areas can also be great for tropical and subtropical bulbs that need heat to grow well. They love a spot next to paving that reflects heat toward them. In these situations, they will grow fast, if you can give them enough water.

A trick to get tender, summer-blooming bulbs into early growth is to start some of your bulbs inside in late winter and early spring. You will need a sunny and warm spot in your house, maybe with a windowsill or extra grow lights. A table near a radiator or on a heat mat can also spur them into breaking dormancy. I love to do this, but for a few weeks my kitchen counters and windowsills do resemble a miniature plant factory. To me, this is worth it. I get my dahlias, cannas, or caladiums to be in full swing a few weeks before they would have done, if I had planted directly out into the garden.

THE YEAR IN BULBS

The most amazing thing about growing bulbs is that even in four-season climates, you can have bulbs that provide interest in your garden or containers for many months of the year. Our bulb year will begin with the bulbs of late winter and early spring. These include snowdrops, crocuses, winter aconites, and rock garden irises.

Spring bulb gardens are extremely floriferous. This is the time to adorn your garden with daffodils, hyacinths, fritillaries, scillas, muscari, and more. You are spoiled for choice with a variety of fragrances, shapes, colors, and types.

The shift to late spring comes slowly and there is some overlap from the earlier weeks. You may still have daffodils blooming, and then tulips erupt into full glory. Alliums follow and then the first of a long season of lilies.

By high summer, the tender plants take over the show. Cannas, caladiums, alocasias, and begonias are all perfect bulbs for your garden once the temperatures rise. Later in the summer, there are more lilies in bloom, and dahlias, gladioli, lycoris, and agapanthus add in their flowers.

By autumn, many of the bulbs of summer are still in full swing, but the last set of new bulbs emerges just when you thought that they were all done. These include sternbergias, fall crocuses, and colchicums. Assembling a cyclical procession of flowers is one of the glories of a brilliant bulb garden.

TOP TO BOTTOM

Begin the bulb year with tiny crocuses that you can plant to pop up through the lawn.

Alliums, like these mixed purple cultivars, keep the bulb show going into late spring.

The late summer and early fall garden is enlivened by the flowers of dahlias, like this display at Fuller Gardens in New Hampshire.

The fall garden is still filled with blooming bulbs like *Cyclamen hederifolium* at the Oxford Botanic Garden, Oxford, England.

Naturalized 'Lilac Wonder' tulips enliven this spring woodland.

2

A SEASONAL PARADE OF BULBS

A garden full of bulbs is a garden that embraces the changing of the seasons. Creating a parade of bulbs is one of the supreme pleasures of gardening because it encourages you to live in the moment and enjoy what is in bloom right now. Bulbs put their all into a super burst of flowers, and then move into a period of rest.

The bulbs that are presented in this section include both familiar ones, as well as others that are less widely known and grown. I emphasize bulbs that are easy to grow in a wide range of climates and soils, that are available for purchase, and which produce flowers or leaves that add to the beauty of the seasonal flower garden. Not all bulbs will be equally successful in every garden. Figuring out which ones will succeed in your garden is a fun process. Try different things, ask gardening neighbors what does well for them, and grow more of the bulbs that you like. If you follow this process, you will eventually develop

a bulb garden that you love. The hardest part is having patience!

The bulb gardening year can be broken out into two halves. The first half of the bulb year begins in autumn. At this time, hardy bulbs that survive your winter temperatures in the ground can be planted. These will wait belowground until late winter, spring, or even summer to flower. This includes many of the spring-flowering bulbs that we are familiar with like daffodils, crocuses, and tulips, as well as later bloomers like alliums.

The second major planting time is in spring, after the soil temperatures have warmed up. This is when you plant the bulbs that are not hardy in your climate. Many of these bulbs will be killed by frost and only grow well in warm weather. These bulbs are the highlights of summer gardens with their colorful leaves and brilliant flowers. We look forward to favorites such as dahlias, cannas, gladioli, lilies, and many more. There are some hardy bulbs that can be planted in fall or spring.

I have divided the seasonal parade of bulbs into these two sections. The first section contains the fall-planted, spring-blooming bulbs, in alphabetical order by genus. The second section covers bulbs that are planted in spring or summer to bloom later that season. They are largely bulbs that are not hardy in the ground year-round and will need to be dug up and brought inside for the winter.

Each plant entry explains the height of the bulbs so that you can know where in the flower bed or container you should plant them, and the growing conditions that they require. It also explains their season of bloom, the colors and shapes of their flowers and leaves, and whether they are fragrant. There are suggested species and cultivars, and design ideas for how to use these bulbs in your garden.

This scented, spring-blooming hyacinth is a hardy bulb that is planted in autumn.

Autumn-Planted Bulbs That Bloom from Late Winter Through Spring

Allium

ornamental onion, allium

Full sun // Moist, well-drained soil // Zones 4–9, varies // Height 8–48 in. // Season of interest—late spring through summer

About These Bulbs

The bloom time for most alliums is after the big spring rush of daffodils and tulips, and they are a welcome addition to the garden at this time of year. A few bloom later to extend the season into summer. Purple is the most frequent allium color with others in white, yellow, or light pink.

The most dramatic alliums are the really tall ones that carry their wide-circumference blooms way above the rest of the bed. They have the wow factor that we need in the late spring and early summer garden. Even a few of these beauties will be a talking point in your garden. The foliage of alliums declines when they are in bloom, so insert them in the back or mid-back of your flower bed, or in the middle of a large container.

Some of these bulbs are on invasive plant lists in certain parts of the world. Check local invasive plant listings. In my eastern Pennsylvania garden, I like to grow fun and funky plants, so I tried *Allium* 'Hair'. It was an aggressive disaster that spread rapidly and was hard to eradicate.

Planting and Maintenance

Allium bulbs are easy to plant at a depth of about three times the height of the bulb. They are not particular about the soil, but like most bulbs, a well-drained spot is best. Some of the very tall alliums grow leaves before their flowers are produced. Their foliage begins to die down early in the season but is easy to remove; or it can be disguised by growing these alliums between surrounding perennials. Once alliums finish blooming, deadhead the spent flowerheads to reduce seedling production.

Design Ideas

Once you have chosen your alliums, look at their eventual heights to see where in the flower bed you should plant them. The tiny ones should be nestled at the front while taller ones can be placed deeper in the bed. Their tall slender stems mean that they act as see-around plants and can come up through perennials that are still low, at the time when the alliums flower. Plant these bulbs around perennials with wide growth habits like catmints (*Nepeta*) or perennial salvias (*Salvia*). Bearded irises are also in bloom at a similar time. I have

CLOCKWISE FROM LEFT: *Allium karataviense* // 'Purple Sensation' and other purple alliums at Waterperry Gardens, near Oxford, England. //*Allium* 'Mont Blanc' // Tall alliums have bold rounded flowerheads.

many mid-height and tall alliums growing through border phlox, which bloom later in the year.

Species and Cultivars

Allium karataviense is a short, very early-blooming ornamental onion. It has a distinctive look with wide glaucous leaves and central white-and-pink-tinged pompom-like flowers. This is a great plant for growing in containers or in a rock garden.

Allium hollandicum **'Purple Sensation'** is the allium that probably pops into your head when you think of this genus. This two-to three-foot-tall allium bears ball-shaped purple flowerheads composed of many star-shaped florets. Each inflorescence is held singly on a vertical stem.

Tall alliums like 'Ambassador', 'Gladiator', and 'Globemaster' are bold purple additions to a flower bed. Their large ball-shaped inflorescences make a statement planted singly or in groups. Plant these bulbs under the edges of low-growing perennials that will hide their fading leaves. They vary in height from two to four feet, so place them in beds accordingly. These flowers have a dramatic effect in the garden because their large size and bold shape is unlike anything else blooming in late spring. They bloom at the same time as bearded irises, lupins, and early roses. 'Mont Blanc' and other similar tall white alliums such as 'White Giant' and 'Mount Everest' are great to use with any colors in the flower bed. 'Summer Drummer' is one of the tallest alliums that blooms later than most other tall alliums in early summer. They can grow up to four or five

Allium atropurpureum // Allium cristophii //Allium unifolium

feet. Plant them at the back or mid-back of the border, and, if possible, position them in front of a fence in case the stem needs support. The flower color is a purple, washed with pale green.

Allium atropurpureum is an atypical allium because its flowers are dark maroon in color with dark green centers to each floret. The bloom looks tightly packed into a hemispherical shape with a flat base. A similarly shaped allium that brings dark maroon to the flower bed is 'Miami'. 'Silver Spring' has blooms that start out light pink and then fade to a creamy white. The center of each starry individual flower begins green and then becomes dark and glossy. 'Pink Jewel' is soft pink with a hint of lilac-and-green centers to each floret. It has a bold presence and a strong stem.

Allium cristophii has a distinct look, beginning as a tight, fat bud that slowly explodes into a loose firework-like ball of starry silver-purple. Position these alliums at the mid-front of a flower bed because everyone wants to look at and touch their blooms. As the flowers fade to beige, they can be dried for winter decorations.

Allium unifolium is native to the Pacific Northwest. It needs a position in full sun with relatively moist soil. It is just over a foot tall, so it is best placed near the front of a bed. They are grown for their clusters of pink flowers that are shaped like stars in late spring. *A. canadense* is native to eastern and central North America. The summer blooms are white and circular on a mid-height plant that blends well with rudbeckia (*Rudbeckia*), penstemons (*Penstemon*), and salvias (*Salvia*).

Allium cowanii // Allium moly // Allium schoenoprasum

Allium neapolitanum Cowanii Group has loose groupings of little white starry-shaped flowers. It is a pretty and easy flower to mix into the front of the bed or to use for cutting. *A. oreophilum* is another small-scaled plant that blooms in late spring with little star-shaped pink blooms held in a loose umbel. Use these plants to come up among violets or another low-growing groundcover.

Allium moly has bright yellow, star-shaped flowers that bloom in late spring. Plant these bulbs at the front of a bed where you can admire the vibrant blooms. They need excellent drainage, and their lax habit means that they look great draped over a rock or bed edging.

Allium schubertii has a foot-wide, open, airy flower with exaggerated distinct, pinky purple florets on long stems, which look like nothing else in the garden. The strange appearance of the bloom comes from the fact that the stalks are all different lengths. The main stem is relatively short, so plant them near the front of the border so that you can see the details of the fun flowers. It makes a great container plant because of the unusual shape. Try one in its own pot.

Allium schoenoprasum, chives, are prized for their onion-like flavor by gardeners who cook. The edible leaves and purple flowers are a tasty addition to dishes. Chives planted in a row make a great edging plant in an herb garden or planted in a container. *A. tuberosum*, garlic chives, is one of the most prolific edible alliums. It is often grown in herb gardens for the plethora of white flowers in late summer. Deadhead promptly to control its rampant spread.

Allium sphaerocephalon// Allium siculum // Allium tripedale

Allium sphaerocephalon blooms in summer. These slender alliums move in the breeze on flexible stems. The flowerhead is a compact ovoid shape in two-tone purple and maroon. This popular plant looks lovely among grasses and perennials such as red hot poker (*Kniphofia*), and sea holly (*Eryngium*).

Allium siculum may be listed for sale by its old name, *Nectaroscordum*. It produces very tall stems topped by pointed buds that pop open to reveal clusters of neutral-toned, bell-shaped florets. They give fabulous height to a flower bed or rock garden area in an unusual, washed, beige-pink color. For best visibility, plant the bulbs in front of a plain background such as a fence, wall, or hedge.

Allium tripedale is similar to *A. siculum* and may be listed for sale as *Nectaroscordum tripedale*. The flowers consist of a cluster of light and dark pink bells with obvious green centers which are easy to integrate with other spring bulbs. Their long flexible stems add vertical interest to the flower bed.

Anemone

anemone

Full sun to Part Shade // Moist, well-drained soil // Zones 4–10, varies// Height 3–18 in. // Season of interest—spring

About These Bulbs

Anemone is a large genus with some lovely spring and early summer bloomers. The short spring-blooming anemones are like miniature stars, which provide a great shape contrast to the other bulbs that are out at the same time. The plants vary greatly in hardiness from the cold-hardy *A. blanda* to the tender *A. coronaria*. The genus includes some species that do not grow from bulbs.

Planting and Maintenance

Anemone tubers and rhizomes need soaking before planting. They all do best in well-drained soil in part to full sun. The plants go dormant after blooming, so plan for something else to grow in that space later.

Design Ideas

Anemones can be grown in containers, naturalized in thin lawn grass, or used as a tiny treasure at the front of a flower bed. They are a good mixer among other plants because their bulbs are small and take up very little space. Their low-growing, ferny-looking foliage does not interfere with neighboring bulbs. Small anemones look good in a rock garden. *Anemone coronaria* is often used as a flower for cutting.

Anemone blanda 'Blue Shades' // Anemone blanda 'Charmer'

Species and Cultivars

Anemone blanda, blanda anemone or Grecian windflower, is the earliest anemone to bloom. Plant the knobby bulbs in part shade in moist, well-drained soil. Plants are short and sweet, rising only four to six inches tall. They make a great temporary groundcover in a semi-shaded place, interplanted with other perennials that fill in after the anemones have died down.

Anemone blanda can be purchased in mixed pastel colors, but it looks more elegant when planted as a single color in one area to coordinate with surrounding flowers. The divided leaves make a good accompaniment to the elongated leaves of other spring bulbs. This is one of the hardiest bulbous anemones, growing in zones 5 to 9.

'Blue Shades' is a subtly colored mixture of light periwinkle blues. 'Charmer' has a distinctive flower with pinkish violet at the ends of the petals, surrounding a ring of white. 'White Splendour' is pure white with a distinct yellow-and-green center. *Anemone apennina* looks similar to *A. blanda* but is taller and blooms slightly later.

Anemone coronaria, poppy anemone, comes in eye-catching colors like red, purple, white, and fuchsia-pink. Poppy anemones are taller and have larger flowers than other anemones. They have a distinctive appearance that is popular for garden display, and as a cut flower. Some cultivars have a black boss of stamens in the center, which may be surrounded by a white or black ring.

In zones 7 or warmer, these anemones should be hardy outside. Plant them where they will get excellent drainage and full sun. In areas with cold winters, they are treated as a tender

Anemone coronaria 'Sylphide' // *Anemone nemorosa*

bulb and lifted in the fall. If they are borderline hardy in your garden, give them a thick layer of mulch to protect them over winter. In all zones, *Anemone coronaria* makes an excellent potted bulb. After soaking the bulbs for about four hours to plump them up, plant the bulbs a couple of inches deep. Plant several in each pot to make a clustered grouping and store the pots in a protected, frost-free place until spring. The flowers will grow up to 18 inches tall. The leaves die down in the summer and the plants should not be watered while they are dormant.

De Caen hybrids are reliable favorites for their single blooms. Try the St. Brigid hybrids for semi-double flowers. 'The Bride' has white single flowers with golden yellow stamens and a green center. 'Mr. Fokker' is dark violet-blue with a black center. 'Sylphide' is a fantastic fuchsia-pink with a prominent ring of black anthers arranged around the central black ball of stigmas.

Anemone nemorosa, wood anemone, thrives in deciduous woodlands among the tree roots, which absorb summer moisture while the anemones are dormant. The flowers are often white, but some lovely selections have light crystalline lavender-blue flowers like *A. nemorosa* 'Robinsoniana'. I grow them in soil heavily amended with leaf mold on a slightly raised bank with cyclamen, snowdrops, epimediums, and lungwort (*Pulmonaria*). Wood anemone grows well in zones 4 to 9.

Camassia

camas, camassia

Full sun // Moist soil // Zones 4–9 // Height 15–36 in. // Season of interest—late spring through early summer

About These Bulbs

These delightful plants have spikes of starry flowers arrayed up their vertical stalks, with separate linear leaves. They are often planted in masses or clumps for impact. Camassias' superpower is that they can tolerate wet soil for most of the year and thrive. These bulbs are primarily native to wet meadows of western North America, where they can form large swaths across the landscape. *Camassia scilloides* and *C. angusta* are a couple of lesser-known species that are native to the eastern and middle parts of North America.

In the garden, camassias are perfect plants to add to a flower bed or wild area for late spring blooms. They grow well in meadow-like settings that are similar to their native habitats. Grassy-looking foliage, which is held low down on the plant, sets off the late spring to early summer flowers that open in sequence, from the bottom of the stem to the top. Their flowers are usually light blue-lilac or dark blue, but some cultivars have creamy white or light pink flowers. An added advantage of these bulbs is that they are deer and rodent resistant. Camassias make good cut flowers.

Planting and Maintenance

Plant camassias where they will receive good light and air circulation. They are one of a handful of garden bulbs that do well in damp areas, such as a rain garden or beneath a downspout. Ideally the soil should be wet in spring, when they are leading up to flowering, and then dry in summer during dormancy. They are perfect for naturalizing in a wild area of your garden, where they can develop into large clumps, by seeding in, and producing offsets from the original bulb.

Design Ideas

Choose your camassias by their height, flower color, and form. The slender flower stalks are sometimes hard to see against the green background of the landscape, especially those with blue flowers. To increase visibility, cluster them together for impact. For a naturalistic look, create groups that contain a different number of bulbs in each cluster and arrange them in organic patterns, avoiding straight lines. They are perfect companions for other plants of damp grasslands, like liatris, meadowsweet (*Filipendula*), and turtlehead (*Chelone*).

Camassia cusickii // Camassia leichtlinii

Species and Cultivars

Camassia cusickii, camas, is a hardy bulb with starry flowers in mid or light blue. When planted in a grouping, they provide a soft, hazy look. The flowers are borne on narrow-stemmed, two- to three-foot-tall plants. It blooms in late spring and is great in zones 4 to 9.

Camassia leichtlinii, giant camas, is the most popular garden camas and has a wide range of cultivars in various colors with either single or double flowers. The plant, including the flower, is 24 to 40 inches tall. It is hardy in zones 4 to 9. Cultivars are plentiful, including the popular long-blooming 'Semiplena', which has tall spikes of double creamy white flowers that show up well in a garden setting. The single white flowers of 'Alba' are starry and elegant with their yellow anthers. The cultivar 'Sacajawea' also has white flowers with a slight white stripe on the foliage. For blue blooms, look for 'Caerulea', a reliable old cultivar with purply blue flowers, or 'Blue Heaven', which is a soft light blue. There is a pink cultivar called 'Pink Star'.

Camassia leichtlinii 'Alba' // Camassia leichtlinii 'Pink Star' // Camassia quamash

Camassia quamash, quamash, is the shortest of the garden varieties at just over a foot. It is good for naturalizing in grass because it prolifically produces offsets. The petals on the individual blue flowers have a slightly twisted look. It is hardy in zones 4 to 9. Two popular cultivars are 'Blue Melody', which has dark blue flowers and leaves that are edged in creamy yellow, and 'Orion', which has green foliage with purply blue blooms.

Crocus

crocus

Full sun to part sun // Well-drained soil // Zones 3–8 // Height 2–6 in. // Season of interest—late winter, spring, and fall

About These Bulbs

Some crocuses bloom in late winter, others in early spring, and a third group flowers in fall. They are all listed here because their shapes, colors, and methods of culture are similar. There is a wide variety of crocuses to choose from. I tend to try a few new ones each year and buy some more of the ones that I would miss if they were not in my garden in the spring.

Spring crocuses are some of the earliest bulbs to bloom in the late winter and early spring garden. Their charming flowers are miniature treats in containers, in window boxes, and at the front of well-drained flower beds. One flower alone is not impressive unless you are down on your knees looking closely, but they are spectacular when clustered together. Their little stemless flowers are goblet- or chalice-shaped and bloom in purple, violet, lavender, gold, yellow, cream, or white. Some crocuses have several colors, often darker at the base and sometimes with stripes or feathered patterns.

Their flowers open wide on a sunny day and close in dull weather. This is a wonderful adaptation because it keeps the reproductive structures carefully protected from snow, or rain, that would otherwise spoil the flower. Crocuses are pollinator favorites at a time of year when few other flowers are in bloom. You will see little bees visiting your flowers on a sunny day when the temperatures warm slightly. Some flowers have a soft honey-like scent. The leaves of crocuses are slender and grass-like, with a thin white stripe that runs along the middle. The leaves continue to elongate after the flowers have faded, to collect sunlight energy for next year's growth. A few weeks after flowering, the spring crocuses become dormant. The fall crocuses have a different lifecycle. They flower in late summer into early autumn, and often their leaves are visible throughout the winter before fading away.

Planting and Maintenance

Crocuses are planted as tiny bulbs that are botanically corms. All crocuses need free-draining soil in a sunny spot. If in doubt, add extra grit or gravel to the soil. Plant crocuses in groups by digging one hole and adding a layer of small-sized grit in the bottom to improve drainage. The hole can be any shape. Sometimes, an elongated swoop-shape close to the front of a flower bed gives the best display. When deciding which way up to plant the crocus, look for the miniature shoot that indicates the top of the bulb, and the flat side, the bottom. Plant the small bulbs, of species crocuses about three inches deep; plant the large Dutch crocus bulbs an inch deeper. Crocuses grow well in pots as long as the soil drains well.

Crocuses are a favored food for burrowing animals and aboveground herbivores. To stop squirrels and other animals from digging up your newly planted crocuses, lay some wire mesh over the surface of the soil. You can remove the wire in the spring or leave it in place and cover it with a layer of mulch. The crocuses will grow up through the holes. If you want extra protection, sprinkle some natural animal deterrent over the soil surface. Division of established crocus clumps is easy. Dig up the crocus corms after flowering and break some off to plant elsewhere. If you have a pest problem, growing them in containers is one of the best ways to protect them.

Design Ideas

Spring crocuses bring one of the first bursts of flower color in the garden. Even the smallest garden has room for a few crocuses. Plant them in well-drained soil in pots, raised beds, or in lawns with sparse unfertilized grass.

Crocuses are tiny treasures that need to be tucked at the front of the flower bed, or close to a path, so that you will see them. They are great on a dry bank, in a trough, or a rock garden. The early bloomers look lovely near deciduous trees and shrubs, which leaf out after the crocuses bloom. I like to arrange my crocuses in groups of one type per hole, though some mixtures of Dutch crocuses can create a cheerful spring show.

In my garden, crocuses are not long-lived because they are often eaten. To hide them, I tuck them against a stone border edging for protection. Alternately, try them among a groundcover like a slender-leaved sedge (*Carex*), lamb's ears (*Stachys*), or lady's mantle (*Alchemilla*). The roots of the perennial plants help keep the crocuses in place during winter rains. They also absorb lots of moisture which keeps the dormant bulbs dry. I like to plant crocuses near plants that have a strong smell like catmints (*Nepeta*), lavender (*Lavandula*), or other herbal plants. Another benefit to this style of interplanting is that you have several seasons of interest in the same part of your bed. Fall crocuses can be added in as well.

Crocus bulbs are ideal for planting in a container. They look fabulous in a small pot. Place them where you can look down into the flowers and smell their subtle honey fragrance. If you want to have a good show, it is best to order new ones each year. The ones that have just finished blooming can be planted into a bed. For a welcoming pot outside your door, combine crocuses with other spring-flowering bulbs and cool-season annuals. Place the crocuses around the edges with some little violas or English daisies. The large bulbs of hyacinths, tulips, or daffodils that produce taller flowers can be positioned at the center or back of the pot. Try growing an all-white spring pot with white crocuses among other little spring flowers, like white scillas, cyclamen, and ipheions. I exclusively grow fall-blooming crocuses in pots of one type per pot. They are placed along my porch wall to greet visitors.

Crocus chrysanthus 'Dorothy' // *Crocus minimus* // *Crocus* 'Ruby Giant'

Species and Cultivars

Early-blooming crocuses bring bright patches of color to the late winter garden. One of the earliest to bloom in my garden is the bright yellow-and-bronze *Crocus ancyrensis* 'Golden Bunch'. Another early one is the Scotch crocus, *C. biflorus*. This diminutive beauty has two white flowers with a blue or purple base and a few stripes. The cultivar 'Blue Pearl' has lilac-blue tips and a lovely yellow throat. *C. chrysanthus*, known as snow crocus, is another early bloomer, sometimes emerging and flowering just as the snow is melting. Cultivars include 'Cream Beauty', which is creamy colored and flushed inside with yellow, or the bright golden yellow 'Dorothy'.

There are some other pretty early spring crocuses to choose from. 'Advance' has creamy petals with a yellow center. 'Ladykiller' has three petals that look painted with dark purple outside with a white upper margin. 'Snow Bunting' has whitish blue outer petals with a touch of yellow at the base. And the rich purple 'Ruby Giant' is sterile and larger than others.

Other species crocuses bloom slightly later. *Crocus etruscus* 'Zwanenburg', the Tuscan crocus, is lilac-purple with stripes outside and orange anthers. *C. minimus* 'Spring Beauty' is a beautiful small crocus that blooms in mid-spring. When you look at the flower from the outside when it is closed, the three outer petals have dark purple vertical feathers on silver-white. Once the flower opens, you can see the purple of the inner three petals as well as the prominent orange anthers. *C. sieberi* 'Firefly' has lilac petals and an orangey yellow center. *C. sublimis* 'Tricolor', one of the most recognizable crocuses, has a yellow

Crocus tommasinianus 'Roseus' // *Crocus vernus* 'Jeanne d'Arc' // *Crocus vernus* 'Vanguard'

bull's-eye surrounded by a white ring and light purple at the top.

Crocus tommasinianus is a slender early crocus that produces an ethereal effect when grown in a mass planting. Known as "tommies," they are often planted to come up through thin turf. They set prolific amounts of seed which makes them a great naturalizer. Squirrels seem to bother them less than other types of crocuses, after their first year in the ground. Some of the nicest cultivars include the warm-toned mauve 'Roseus'; 'Whitewell Purple', which is reddish purple with a pale center; and the darker lilac-purple 'Barr's Purple'.

OPPOSITE: *Crocus biflorus* 'Blue Pearl'

Crocus vernus, Dutch crocus, is larger and later to bloom than most spring-blooming species. Dutch crocuses are the biggest and brightest of the crocus bunch. They are available as single colors (including brilliant egg-yolk-yellow, bold lilac-purple, and creamy white) or mixtures. Dutch crocuses grow well through chipped gravel or along the front edge of a flower bed. They are excellent in containers, either when grown alone, or combined with coordinating early-blooming daffodils.

I particularly like the cultivars 'Flower Record', with dark purple flowers and prominent orange stamens, 'Jeanne d'Arc', with lovely petals of white adorned with faint purple lines, 'Pickwick', a good grower with a pinstripe pattern, and 'Vanguard',

Crocus speciosus // Crocus 'Zephyr'

with an attractive two-tone effect that looks white when closed, and purple when open.

Autumn-blooming crocuses are spring-looking flowers that surprise you when they emerge in late summer or early fall. *Crocus speciosus* has translucent lavender-blue petals and a divided orange stigma. The cultivar 'Oxonian' has darker petals. *C. kotschyanus* has lilac flowers with yellow centers. *Crocus* 'Zephyr' has silvery, light blue flowers. *C. sativus* is a great garden plant that is also the source of the spice saffron. It has beautifully striped petals and distinctive long stigmas, which have been grown as a crop for centuries. To harvest the stigmas for cooking, pick and dry them. They can be stored in an airtight glass jar. Before you eat anything from your garden, make sure that it is correctly identified.

Cyclamen

cyclamen

Part shade // Moist, well-drained soil // Zones 5–9 // Height 4–6 in. // Season of interest—late winter through late summer

About These Bulbs

Hardy cyclamen are grown for their blooms that light up areas of autumn or spring sunshine. They are short in stature but can spread to create a charming picture. Their delightful leaves are decorative features that outlast the flowers. Some have plain green leaves, but those with silver sheens or mottling are highly desired. You can grow large-flowered, non-hardy cyclamen in pots, window boxes, or indoors for winter bloom. Cyclamen are deer resistant.

Their petite five-petalled flowers are reflexed and sweep upward at the tips. The bloom colors range from bright pink to nearly red, purply pink, light pink, and white. The flower stalks arise in a circle from around the flat circular bulb. Some cyclamen flowers have a light fragrance. Cyclamen may self-seed if your growing conditions are right. When the flowers are pollinated, seed capsules develop on spring-like structures. When the seed is ripe, they uncoil down to the soil to release the seeds.

Planting and Maintenance

Plant the flat disc-like bulb at, or just below, the surface in a woodsy, free-draining soil that does not dry out completely during dormancy. The flat or concave side is the top and the rounded or convex side is the bottom. Wherever you plant them, make sure that they are not where you might dig them up later in the year. Once the top growth dies down, you will not be able to see where they are planted.

Improve the soil by adding some small-scale grit and leaf mold to the planting hole, and as a top-dressing. The shoots come up around the center of the bulb so they should not be buried too deeply or mulched heavily. Cyclamen are drought tolerant once established. Do not overwater, especially during dormancy. To increase drainage in the planting area, raise the planting bed by adding large stones at the front. The bulbs do well when tucked next to rocks where drainage is good, such as in a partly shaded rock garden. If planting cyclamen in containers, ensure that the soil has great drainage, and keep them evenly watered.

My hardy cyclamen are planted on berms that are sheltered by deciduous trees and shrubs. The cyclamen plants benefit from the early spring sunshine, followed by protection from strong sun by the woody plants. The tree roots help the survival of the cyclamen by removing moisture from the soil in summer. An occasional sprinkling of leaf mold keeps the soil moist, yet well-drained.

Design Ideas

Plant cyclamen along an accessible path so that you can see the flowers easily. Hardy cyclamen look great naturalized near deciduous trees and shrubs. Plant *Cyclamen coum* in conjunction with snowdrops, crocuses, lungwort (*Pulmonaria*), and winter-blooming witch hazels (*Hamamelis*) for an early spring scene. *C. hederifolium* looks great as a mass carpet by itself or with colchicum. It is a larger plant and can outcompete the smaller *C. coum*, if they are planted in the same space.

Florist's cyclamen are wonderful in window boxes and containers. Their leaves are delightful up close. Choose white-flowered cyclamen with silver leaf markings for a frosty look with white winter pansies or choose pink or red flowers for a bright pot. They also make a great indoor-blooming plant for the winter months.

Species and Cultivars

Hardy cyclamen species include the fall-blooming *Cyclamen hederifolium*, and the late winter and early spring *C. coum*. They are both hardy in zones 5 to 9. *C. coum*, spring-blooming hardy cyclamen, has petite flowers in shades of pink and white that emerge in late winter or early spring, depending on the severity of your cold season. It has rounded kidney- or heart-shaped leaves, often with silver- or pewter-colored variegation. The foliage emerges in late autumn and stays green for the winter. *C. hederifolium*, the ivy-leaved cyclamen, has mottled trilobed leaves that resemble those of an ivy plant (*Hedera helix*). This cyclamen blooms in late summer into early autumn. Their leaves emerge in the fall and last throughout the winter. Flowers may be white, pink, or magenta.

Cyclamen persicum, Persian cyclamen, as well as its cultivars and hybrids, called florist's cyclamen, can be used inside as a houseplant for their winter blooms, or planted outside where they are hardy in zones 9 to 11. To keep the cyclamen plants for next winter, store the bulbs indoors after they have flowered. Keep them relatively dry before planting them up again in autumn, when you should start regular watering. These are often purchased as growing plants rather than as bulbs.

TOP: *Cyclamen coum* // BOTTOM: *Cyclamen hederifolium*

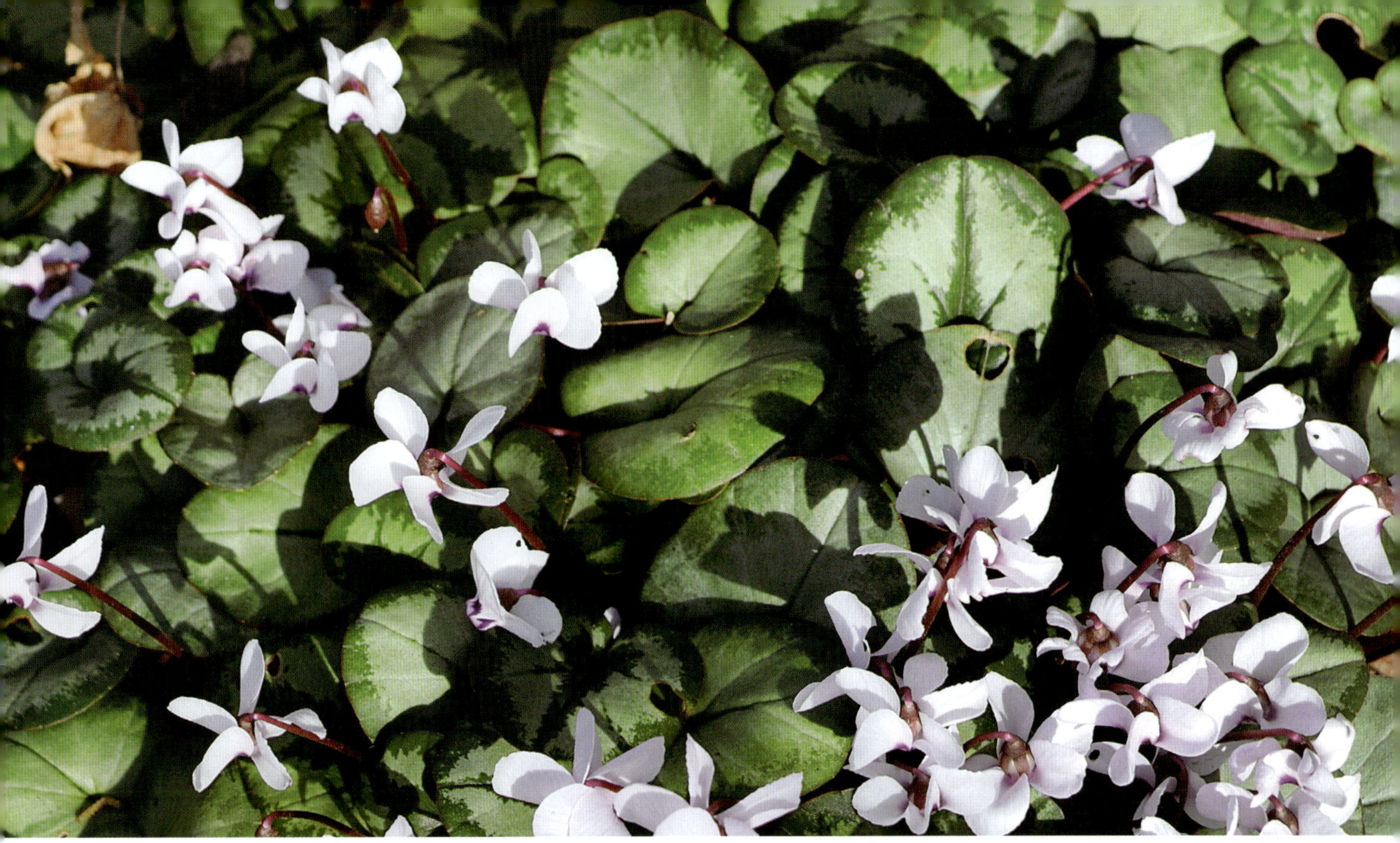
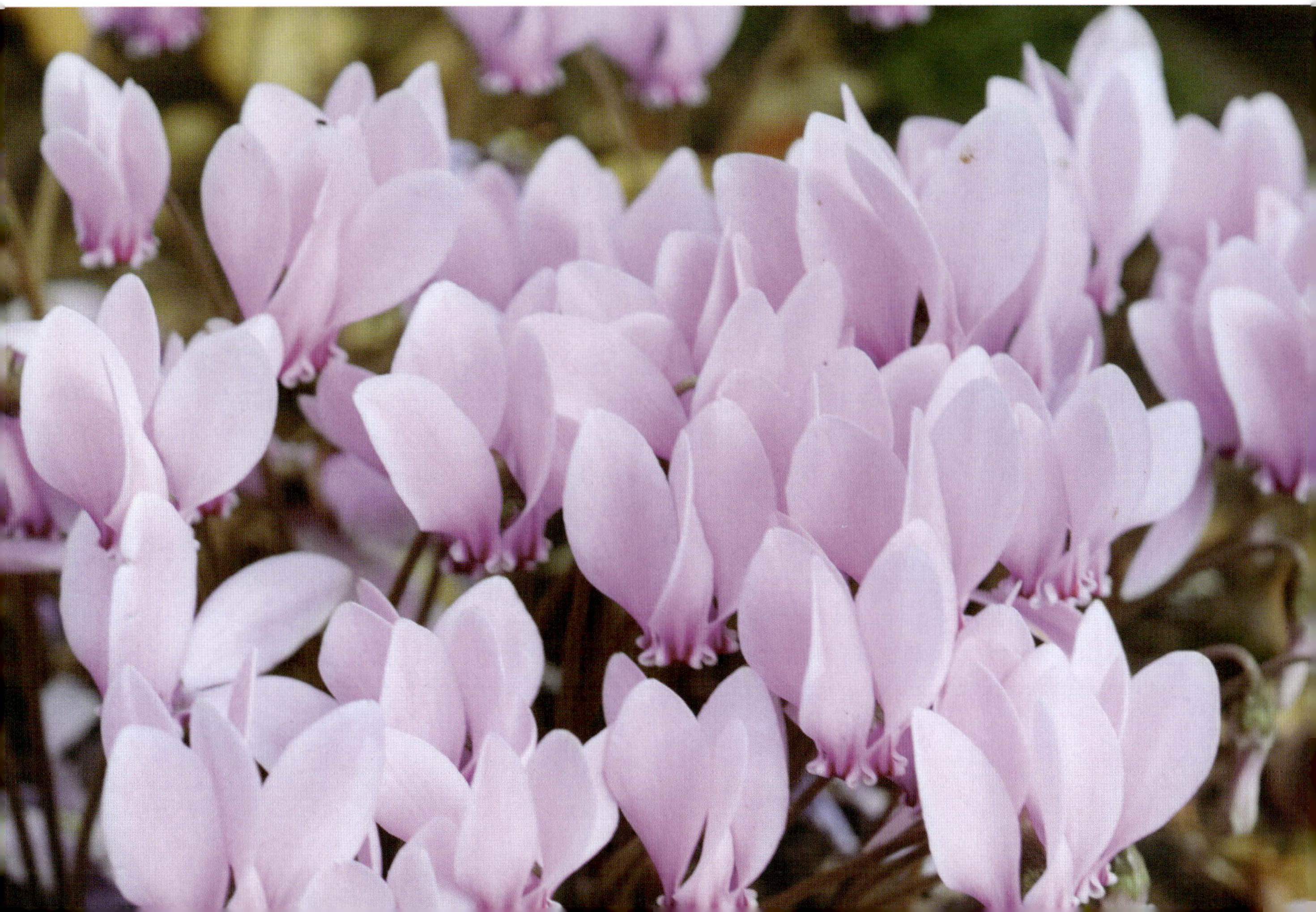

Eranthis

winter aconite, eranthis

Part shade // Moist, well-drained soil // Zones 3–9 //
Height 2–4 in. // Season of interest—late winter

About These Bulbs

Eranthis hyemalis, winter aconite, is a cheerful late winter and early spring bloomer, which may emerge while snow and frost are still lingering. Winter aconites are perfect for naturalizing around trees, shrubs, and roses, or on slopes. Their shiny, bright yellow, buttercup-like flowers face upward on short stems and attract winter pollinators. In the center of the petals is a yellow tuft of stamens and around the outside is a ruff-like green bract. The leaves are a prominent feature after the flowers fade. They act as a temporary groundcover before dying down late in spring. Winter aconites are not bothered by deer or rabbits.

Planting and Maintenance

Plant bulbs a couple of inches deep in rich, woodsy soil under deciduous trees, where there is full sunlight while the winter aconites are in bloom and then shade later on. Dried-up bulbs should be soaked in water overnight to rehydrate them before planting. A scoop of actively growing plants and their tubers will thrive if quickly transferred to a new position in spring. You can also use this method to make more clumps in your own garden.

Eranthis can be hard to get established, but if they seed themselves into your garden, you may find that they are overly aggressive. These bulbs will grow in a range of soils including alkaline soil.

Design Ideas

Eranthis hyemalis is one of the first bulbous flowers to bloom in the seasonal bulb calendar. Plant them close to your house where their bright yellow blooms will be visible, no matter how bad the winter weather. For a lovely late winter show, grow them intermingled with snowdrops and crocuses.

Species and Cultivars

Eranthis hyemalis, winter aconite, is the species that is usually available. It has a few cultivars, but they can be hard to source. 'Orange Glow' has an unusually rich and warm color. *E. hyemalis* Cilicica Group is a lesser-known winter aconite that has a bronze cast to the emerging leaves; its flowers bloom a little later than straight species winter aconite.

TOP: *Eranthis hyemalis* // BOTTOM: *Eranthis hyemalis*

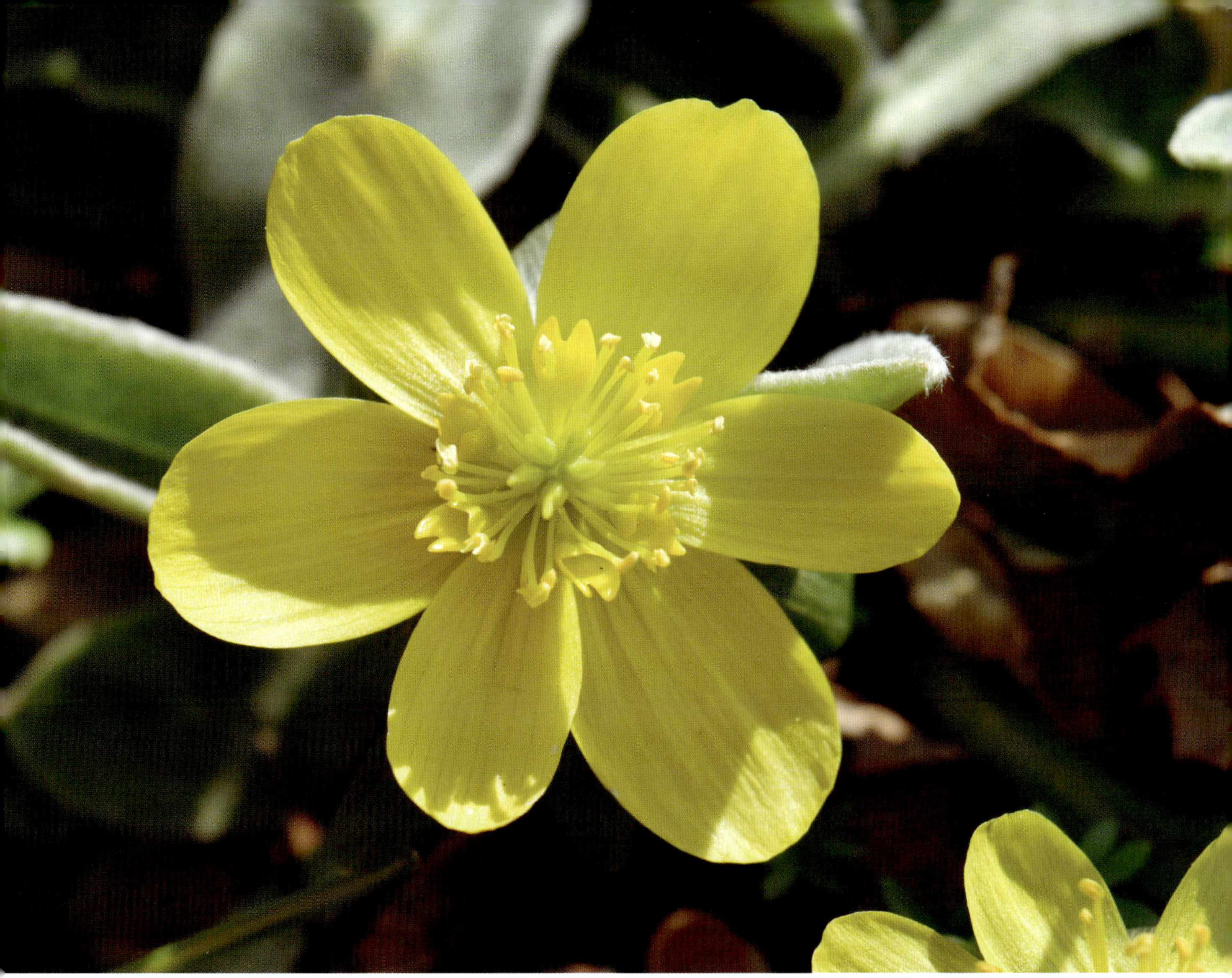

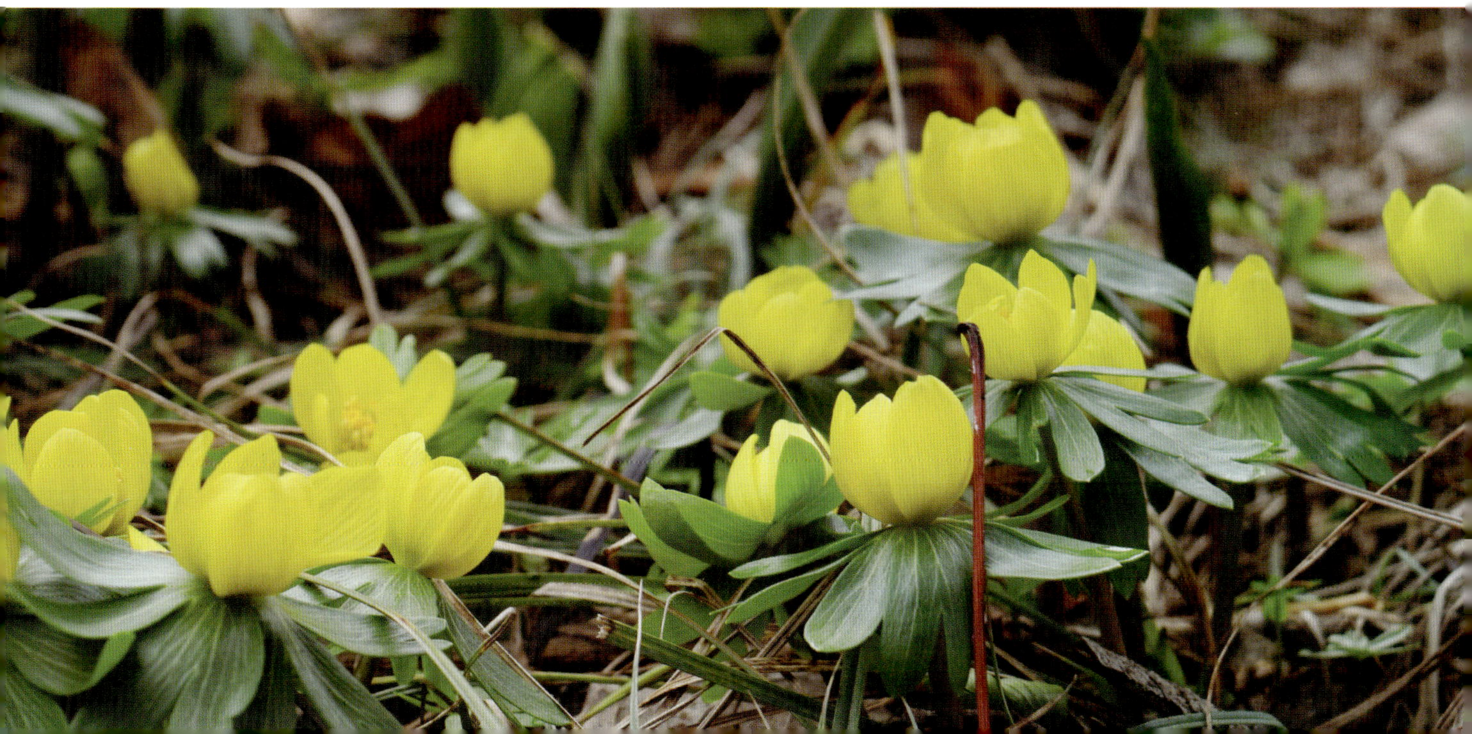

Eremurus

foxtail lily, desert candle, eremurus

Full sun // Rich, well-drained soil // Zones 5–8 // Height 36–90 in. // Season of interest—late spring through summer

About These Bulbs

Foxtail lilies are fun statement flowers that tower over the spring and early summer floral displays like exclamation marks. They may be three to six feet tall, with some species reaching eight or nine feet. Hundreds of little six-petalled star flowers open at the base, and then a wave of bloom ascends to the top of the spikes. Foxtail lilies range in color from white, cream, and yellow to blush, peach, and coral orange. Their protruding anthers make an added decorative statement during their long bloom time and attract lots of small insects to the pollen. Foxtail lilies can be used as a special cut flower in tall flower arrangements. They are deer and rodent resistant.

Planting and Maintenance

These plants grow from large and strange-looking spidery bulbs. There is a central bud called a crown, which is surrounded by long, fat, root-like structures. Plant the crown two to four inches deep.

If you garden in sandy soil, plant them slightly deeper. Handle these fragile bulbs with care. Leave them undisturbed once they are planted.

Each bulb will need a large hole to itself—some bulbs can be the size of dinner plates. Find a sunny site to plant them, where they will also be sheltered from the wind. They are tall plants that can get blown over. A situation at the back of a border near a fence is ideal.

Eremurus needs rich soil that is high in organic matter, but also well-drained. The leaves form a basal rosette that appears in spring. The emerging shoots are sensitive to frost, so protect them with a layer of leaf mold or other mulch. The plants grow rapidly, so keep the roots well watered to help them achieve their maximum growth rate. After flowering, cut the tall stem back to the ground and let the foliage die down naturally. The bulb needs low water throughout the summer, fall, and winter. Wet soil, especially over winter, will kill this plant. I have planted them on a sunny raised bank that is mulched with a couple of inches of gravel.

Melon-colored eremurus flowers up close. // Eremurus flower spikes rise above lower plantings.

Design Ideas

Foxtail lilies are prominent vertical features when they are in bloom, so they will be noticed wherever you plant them. Their bulbs take up a lot of room below the soil. However, the actual spike is slender, which leaves room to scatter annual flower seeds around the stem. Another idea is to plant them among tall perennials at the back of the bed. These later-blooming perennials will fill in and use the same area that the foxtail lilies occupied earlier in the growing season. Positioning their pretty flowers in front of a plain background, like an evergreen hedge or fence, will make them visible in the garden.

Species and Cultivars

Eremurus robustus, giant desert candle, is very tall, growing to eight feet with flowers in various pale pink shades. *E. stenophyllus* has light yellow flowers on three- to five-foot-tall stems. It is more common to find hybrids for sale listed as *E. ×isabellinus*. Choose the ones you want to grow by looking at their ultimate heights and their colors. Some of the finest are the five-foot-tall 'Cleopatra', which has dark orange buds that open to orange flowers, and 'Orange Marmalade', with coral-colored flowers. Other hybrids include the four- to five-foot-tall 'Romance', with flowers in peachy pink, and 'White Beauty', which has four-foot-high stems and creamy white flowers.

Erythronium
trout lily, fawn lily

Part sun // Rich, moist, woodsy soil // Zones 3–9 //
Height 6–12 in. // Season of interest—spring

About These Bulbs

Erythroniums are fantastic spring bulbs for a moist, partly shaded woodland area. Trout lilies, or fawn lilies, have wonderful glossy and often patterned foliage. These markings are thought to resemble those on the animals used in their common names. Their delicate reflexed flowers hang from flexible stems. These bulbs are primarily native to parts of North America, where their natural habitat is moist forests and partly shaded meadows. A few species are native to sections of Asia and Europe.

Planting and Maintenance

Plant the bulbs with their pointed ends up about four inches deep in moist, well-drained soil immediately on receipt—these bulbs resent being out of the ground. Protect the bulbs from rodent predation and from drying out even when they are dormant. As plants of wooded areas, they benefit from a mulch of leaf mold.

Design Ideas

Use trout lilies in a naturalistic woodland setting with other spring bulbs like crocuses, snowdrops, and shade-tolerant species tulips. Interplant the bulbs with primroses (*Primula*), hostas (*Hosta*), and ferns that will expand into the space left after the trout lilies die down.

Species and Cultivars

The most popular *Erythronium* is the hybrid cultivar 'Pagoda', which grows well in garden settings. It has slightly mottled leaves and dangling yellow bell-shaped flowers. Other available trout lilies include the cute, white-flowered *E. californicum* 'White Beauty'; yellow *E. rostratum*; and the cultivars of *E. dens-canis*, such as the purply 'Lilac Wonder'.

OPPOSITE CLOCKWISE: *Erythronium* 'Pagoda' // *Erythronium californicum* 'White Beauty' // *Erythronium rostratum*

Fritillaria

fritillary

Full sun to part shade // Well-drained or moist soil //
Zones 4–9 // Height 6–48 in. // Season of interest—spring

About These Bulbs

Fritillaria is a wonderful diverse genus of bulbs that includes some really good garden plants. There is something for every area, from tiny treasures for the front of the bed, to towering bold beauties for the mid-back of your plantings. Whichever fritillary you choose, it will become a talking point due to its intriguing beauty. I am rather obsessed with this group. They are distributed around my garden, so that I can place them where they will grow best.

The coloring and shading of fritillary flowers are rather unusual and range widely, from subtle to brassy and brazen hues. Some have a checkered pattern and others a gray-green wash over the flower. If you love moody colors in your flower bed, then include some fritillaries. These plants are not eaten by rabbits, deer, or other burrowing animals. Some of the bulbs, particularly the crown imperial, have an odor that is similar to a skunk or a fox.

Planting and Maintenance

Most fritillaries need to be planted in a raised position that provides great drainage. Plant them away from irrigation systems, otherwise they will rot. Choose a place in full sun for most fritillaries. *Fritillaria meleagris* does well in moist soil and can take part shade; *F. pallidiflora* can also handle light shade.

Plant the bulbs as soon as you receive them. They do not have a protective outer coating and can desiccate while they are out of the ground. Dig a hole to the right depth for the size of the bulb. Large bulbs like *Fritillaria imperialis* and *F. persica* need to be planted six to eight inches deep. Plant smaller fritillary bulbs at about five to six inches. The top of each bulb has an indentation, or hole, where the stem emerged last year. When you look at the bulbs, you may be mystified as to which way is up for planting. If you are in doubt, lay them on their sides. The bulbs will eventually pull themselves into the correct position in the soil, using their roots. If your soil is heavy, add extra grit beneath the bulbs to increase drainage.

Let the leaves die down after they flower. Once the leaves have become straw-yellow in color, they can be gently pulled away and composted. If you do not deadhead spent flowers, they may seed in. A thick layer of organic mulch will prevent seeding, whereas gravel mulch is ideal for encouraging self-seeding.

The major pest of fritillaries is the bright red lily beetle. If you see any, remove and destroy them. The easiest way is to squish the beetles, but be aware that they fly away quickly. Slugs and snails can eat fritillaries, especially when the young shoots are emerging. To reduce mollusk hiding places, keep the area around the little plants free of leaves in the spring.

Design Ideas

Choose your fritillaries according to the planting conditions that you have. The short ones do really well in a gravel garden or rock garden where it is easy to admire the intricate details of their flowers. Grow them with low-growing perennials, such as creeping phlox (*Phlox*) or candytuft (*Iberis*), or alongside short bearded irises. Any fritillary can be used in containers or flower beds alongside other spring bulbs like muscari and scillas. The tall fritillaries look dramatic rising up through tulips and tall daffodils. Pick up the warm hues of crown imperial fritillaries by surrounding them with orange or yellow flowers.

Species and Cultivars

Fritillaria imperialis, crown imperial fritillary, is the tallest and most dramatic plant in the genus. Its commanding presence in containers and flower beds is unlike almost anything else in the garden. Strong straight stems rise above the spring garden and are topped with tufts of green bracts that look like leaves. Below these hang rings of red, yellow, or orange dangling bell-shaped flowers. The lance-shaped leaves are typically green, though cultivars may have variegated or dark leaves instead. The stems do not need staking.

'Rubra Maxima' has deep orange flowers paired with strong dark stems that look great when surrounded by dark colored tulips. 'Lutea' has yellow flowers that look good with daffodils and muscari. 'Aureomarginata' has light orange flowers and variegated leaves. 'Red Beauty' has red flowers and green foliage. 'Sunset' bears orange flowers with bronze stems and bracts.

Fritillaria persica, Persian fritillary, is an elegant vertical flower for a fertile, well-drained site. Tall, upright stems hold dangling, bell-shaped flowers that open in sequence from the bottom of the stalk upward. The straight species has dark maroon-slate flowers that are best in a light-colored planting, so that their flowers are visible. 'Ivory Bells' is creamy in color, as its name suggests, with golden anthers; 'Green Dreams' is green that shades to purple-brown; 'Purple

Fritillaria imperialis 'Sunset' // *Fritillaria persica* // *Fritillaria meleagris*

Dynamite' has shiny dark purple flowers that are less bell-like than some cultivars.

Fritillaria meleagris, checkered fritillary, has nodding maroon and cream flowerheads with a subtle checkerboard pattern, that are held singly or doubly, on a foot-tall plant. It makes a charming addition to a damp, partly shaded section of the garden. Try it in an area near a downspout, or around a birdbath that you fill regularly. The white cultivar 'Alba' shows up well in a naturalistic lawn planting.

Fritillaria thunbergii has flowers that are not brightly colored, so plant them in groups for best impact. To see the amazing checkered pattern on the inside of the flowers, tip up the nodding heads. Their unusual leaves may grasp and climb into nearby plants. I grow this fritillary around a pond, with several types of primroses (*Primula*) and blue forget-me-nots (*Myosotis*). They may seed around the garden.

Fritillaria thunbergii // Fritillaria michailovskyi // Fritillaria uva-vulpis

Short- to medium-height fritillaries

(apart from *Fritillaria meleagris* mentioned above) all require well-drained soil. They are perfect when planted in a gravel garden or rocky area, making good companions to rock garden perennials. *F. acmopetala* is a mid-height fritillary with one or more distinctly bell-shaped olive-green, yellow, and maroon flowers that flare out at the rim. Rising only six to eight inches, *F. michailovskyi* is a short bulb for the front of the flower bed. It has maroon-purple bells that are edged with bright yellow. Pair them with miniature daffodils. *F. pallidiflora* has a stem that is a foot or more tall from which hang creamy yellow flowers with tiny purple dots. It thrives in rich, well-draining soil, and light shade. *F. uva-vulpis* is about a foot tall and has little ovoid blooms that dangle on arched stems. The yellow-rimmed maroon flowers are covered with a dusty bloom like you find on dark grapes. When planted in ideal well-drained soil, this fritillary will multiply and make a lovely stand.

Galanthus

snowdrop

Part shade // Moist, well-drained soil // Zones 3–9 // Height 4–12 in. // Season of interest—winter through spring

About These Bulbs

The first bulbs to flower in late winter and early spring are the little white snowdrops. They are a perfect addition to a slightly shaded area of the garden. All snowdrops are primarily white, but with small green or yellow markings. The species and cultivars are differentiated by the size, shape, position of the markings, plant height, and overall form of the flower. The strap-like foliage is sometimes slender and sometimes broad, but still elongated. The leaves can be glaucous green, or bright green and shiny.

Planting and Maintenance

Plant snowdrop bulbs in the fall about three inches deep, into partly shaded, good woodland soil. I have found that the best place to plant them is on a bank or slope near deciduous trees and shrubs, which provide shelter from the worst of the summer heat. Snowdrops are not bothered by rodents, rabbits, or deer. Snowdrops are very amenable to being dug up and divided "in the green," while their leaves are still visible. Snowdrops propagate themselves by seeding in.

The first year that the seeds drop and germinate, the resulting seedlings may look like tiny blades of grass. Be careful not to weed them out at this stage, or to cover the snowdrop bed with too much mulch, if you want seedlings.

Design Ideas

Snowdrops are the perfect late winter and early spring garden treat. They look good tucked into any little partially shaded corner. Try a few at the base of a wall or a hedge, pop a few by the mailbox, and another couple by the door. If you have many snowdrops to spread around, they look lovely in random clumps, lining a path, and coming up through weak lawn.

Interplant your snowdrop clumps with deciduous herbaceous plants that will come up later in the spring and fill the space where the snowdrops were growing earlier. I have found that some of the best companions are lungworts (*Pulmonaria*) and deciduous ferns, such as Japanese painted fern (*Athyrium niponicum* var. *pictum*). The leaves of these other perennials die down in autumn and pop up afresh after the snowdrops have had their moment in the sun.

CLOCKWISE FROM LEFT: *Galanthus elwesii // Galanthus nivalis // Galanthus nivalis* 'S. Arnott' // *Galanthus* 'Richard Ayres'

Snowdrops also look good in pots. Either plant them up in the autumn, or if your ground is not frozen, dig up a few when they are just emerging from the ground. Carefully place the whole clump into a container and top-dress with some moss or gravel.

Species and Cultivars

Galanthus elwesii, giant snowdrop, is the tallest of the garden snowdrops, topping out at about 8 to 12 inches with slightly longer leaves. Giant snowdrops have been the best for me in my mid-Atlantic North American garden. They seem to relish a position on a bank that is bathed in spring sunshine. The earliest giant snowdrops to bloom have one green mark on the inner part of the flower; the later have a pair of green marks one above the other. The leaves are broad, flat straps in a gray-green color.

Galanthus nivalis, common snowdrop, is native to European deciduous woodlands. It naturalizes in dappled shade and can increase and spread so successfully that the sheets of snowdrops resemble a snowy landscape. The single white flowers are touched with green on the inner petals. The fragrance of common snowdrops is like honey and can be smelled on a late winter day, as the sun warms the flowers. 'S. Arnott' is one of the best for vigor, availability, and scent.

Galanthus nivalis 'Flore Pleno', double snowdrop, has a ruffled look. If you tilt up the flower and look at it from the underside, you will be able to see the many layers of petals and

Galanthus woronowii // Galanthus nivalis 'Viridapice' // *Galanthus* 'Blewbury Tart'

some subtle green markings. This snowdrop is a prolific increaser, so make sure that you dig and divide the groupings regularly. Some of the named doubles have very even flowers, like the well-formed 'Richard Ayres'.

Galanthus woronowii was the first snowdrop that I bought for my current garden, so it has a special place in my heart. It flowers later than the other species, with distinctive, glossy, bright green leaves. The flowers are slender and delicate. The overall effect is not so floriferous as a mass of common snowdrops.

Special snowdrop cultivars are one of the treats of the late winter garden. Depending on where you garden, you may be able to find a wide range or only a few types. My advice is to begin with the tried-and-true old-fashioned snowdrops that you know are strong growers. When you find places that snowdrops do well in your garden, then you can buy some of the more expensive and harder to find cultivars. 'Viridapice', which has obviously green tips to the outer petals even when in bud, is held aloft on a long stem. These bulbs are fantastic increasers. There are recognizable ones with yellow ovaries (such as 'Spindlestone Surprise'), ones with upward-facing flowers that are irregular (like 'Blewbury Tart'), and many more.

OPPOSITE: *Galanthus* 'Spindlestone Surprise'

Hyacinthoides

English bluebell, Spanish bluebell

Part shade // Moist, well-drained soil // Zones 5–9 //
Height 12–18 in. // Season of interest—spring

About These Bulbs

Bluebells are grown for their delightful flowers that are borne along the top half of the leafless flower stem. Each flower resembles a miniature bell, in colors of traditional blue-violet, white, or pink. The strap-shaped leaves emerge from the base of the plant. This is a plant of my childhood woodland walks. I remember turning the corner and seeing a sea of light blue, as the understory of deciduous woodlands, and smelling a slightly honeyed scent wafting through the trees.

Planting and Maintenance

Hyacinthoides is wonderful planted among deciduous trees or shrubs. Here, they can collect spring sunshine, and then have summer shade, to keep the bulbs dry but not desiccated. Plant the bulbs about three inches deep in the fall into good humus-rich soil. Arrange the bulbs in clumps by digging one large hole, and placing a few bulbs in there together, spaced out but still looking like they relate to each other. They should multiply over the next few years. After flowering, allow the foliage to die back naturally. Once they are dormant, a leaf mold mulch is a good addition.

Design Ideas

Bluebells grow well with woodland anemones, trout lilies, and snowdrops. They look lovely coming up through a low-lying groundcover, or in between deciduous ferns. They can also grow in mixed flower beds alongside tulips, daffodils, hyacinths, and violets (*Viola*). Try them near a door or cut some flowers so that you can appreciate their scent.

CLOCKWISE FROM TOP LEFT: *Hyacinthoides* look lovely when planted in groups in part shade. // *Hyacinthoides hispanica* // *Hyacinthoides non-scripta*

Species and Cultivars

Hyacinthoides non-scripta, English bluebell, has graceful bell-shaped flowers that hang down from one side of their arching stems. The flowers are a fantastic blue color, with a tinge of light purple. This bluebell has a slight honey scent.

Hyacinthoides hispanica, Spanish bluebell, has larger and bolder-looking flowers than the English bluebell. A variety of cultivars are available in pastel colors of light lavender, blue, light pink, and white. Some of the nicest are 'Dainty Maid' and 'Queen of the Pinks', which have pink flowers; creamy white 'White City'; and 'Excelsior', in shades of blue and blue-purple.

Hyacinthus

hyacinth

Full sun // Rich, well-drained soil // Zones 4–9 // Height 8–12 in. // Season of interest—spring

About These Bulbs

Hyacinths are the best bulbs for spring fragrance. They are easy to grow outside in the ground, in pots, or to force for earlier bloom inside the house. Each flower spike is composed of many bell-shaped individual flowers that flare out into star-shaped openings. Double-flowered hyacinths have a full look. If you want a hyacinth with a graceful appearance, look for the multiheaded hyacinths, which have smaller flowerheads.

Hyacinths are available in a beautiful color range from soft cool white, cream, blue, pink, and salmon to richer hues of dark blue, deep purple, plum, beetroot, and coral-orange. Their wide, green, and slightly shiny leaves are held at the base of the plant.

Planting and Maintenance

Plant hyacinth bulbs in autumn in averagely moist, well-drained garden soil, about five to six inches deep. Ensure that the bulbs get moisture during the growing season but are not wet during the summer or winter. Add grit to the bottom of the planting hole to provide extra drainage. Hyacinths are fabulous in pots, either by themselves or to add color and fragrance to mixed spring containers. Hyacinth bulbs should be handled with gloves, as they may irritate your skin. Wash your hands after planting them.

Newly purchased hyacinth bulbs produce large, dense flowerheads. In subsequent years, the flowerheads become sparser and look more natural. In the ground, I like this softer look. The flowerheads hold themselves up easily. Some gardeners replace their hyacinth bulbs each year, so that they get the full-flowered look each time. With any hyacinths, chop the whole flowering stem off to prevent seed production, and to allow all of the energy to go back to the bulb for next year. Remove the leaves once they turn brown.

Design Ideas

I always include hyacinths in my spring garden designs for their heady scent. Public gardens use massed displays of hyacinths; in a home garden, it only takes a few bulbs tucked into a pot or garden to remind you that spring has arrived. Plant them in a sunny corner where the fragrance can linger in the air. Clustering hyacinths of the same color together has a more cohesive look than mixed colors planted together. I grow hyacinths in small

CLOCKWISE FROM LEFT: White 'Aiolos' and dark 'Blue Sapphire' hyacinths // *Hyacinthus orientalis* 'Splendid Cornelia' // *Hyacinthus orientalis* 'Miss Saigon' // *Hyacinthus orientalis* 'Aqua'

groupings located a foot or two back from a path to a side door, in a position where we can see and smell them often. Pastel-colored hyacinths are combined with tulips like 'Lady Jane', 'Spring Green', and 'Honky Tonk', alongside mid-height daffodils like 'Tristar', and the short, cute 'Oxford Gold' at the front. To disguise the withering foliage, I plant spring-growing perennials like catmint (*Nepeta*) near the bulbs.

Hyacinths look great when grown in pots. Combine them with small-scale daffodils,

hellebores (*Helleborus*), and cool-season annuals like English daisy (*Bellis perennis*). For natural-looking supports, use some pussy willow or red twig dogwood stems inserted into the soil around the hyacinths.

Species and Cultivars

Hyacinthus orientalis, hyacinth, has a wide color range that gives delightful choices when reinforcing a planting scheme. If you are

CLOCKWISE FROM TOP LEFT: *Hyacinthus orientalis* 'Atlas' // *Hyacinthus orientalis* 'City of Haarlem' // *Hyacinthus orientalis* 'Sweet Invitation' // *Hyacinthus orientalis* 'Fondant' // 'White Festival' is a multiflowering hyacinth.

looking for a dark maroon-purple hyacinth, try 'Woodstock'. Other purply cultivars include my favorite 'Splendid Cornelia', which has a blue tinge, and the slightly darker 'Miss Saigon'. Blue hyacinths range from the dark 'Blue Jacket' and 'Aqua', to the lighter old-fashioned favorite 'Delft Blue'. For a white hyacinth, choose 'Aiolos', 'Carnegie', 'Atlas', or 'White Pearl'. 'City of Haarlem' is a good yellowy cream heirloom favorite. For

coral touched with pink, try 'Sweet Invitation'. If you want a light pink, pick 'Fondant' or 'Pink Pearl'. 'Jan Bos' is vibrant reddish pink. There are a few doubles to choose from like 'Royal Navy' or 'King Codro', both of which are very dark blue, with a touch of purple. If you want the smaller-headed hyacinths, look for light blue 'Blue Festival' or fresh white 'White Festival'.

OPPOSITE: 'King Codro' is a double hyacinth.

Ipheion

ipheion, spring starflower

Full sun // Moist, well-drained soil // Zones 5–9 //
Height 3–6 in. // Season of interest—spring

About These Bulbs

Ipheion uniflorum is a lesser-known bulb that deserves a place in more gardens. It produces upward-facing, star-like flowers, with a tiny yellowish eye. The flowers have three inner petals that tilt slightly upward and three lower ones that angle downward. A delicate stripe runs along the back of the petals that is barely visible from the front. The flowers may be soft pastel blue, periwinkle, pink, purple, or white, and they are honey-scented with a touch of spice. The low, grass-like foliage smells rather like onions or garlic when crushed or cut. Consequently, it is deer and rodent resistant. It is also a good pollinator plant.

Planting and Maintenance

Spring starflower bulbs are easy to plant in moist, well-drained garden soil at a depth of about three inches. When you plant them, dig one hole, and make a nice grouping of bulbs that will create a full clump. If your soil is heavy clay, add some extra grit and organic matter to the flower bed at planting time. If you have cold winters, mulch heavily and choose a sheltered microclimate near the warm side of the house. When using them in lawns, make sure that the foliage can ripen and die down before the grass is mowed. The leaves start into growth again in early spring. Ipheion plants are a favorite spring treat for slugs and snails, so check the plants regularly.

Design Ideas

Ipheion is an ideal low-growing plant for the edge of containers or for edging a flower bed. It can be used in rock or gravel gardens or for naturalizing in lawns. Combine it with other small-scale spring bulbs like muscari, crocuses, and scillas for a nice sequence of bloom.

Species and Cultivars

Ipheion uniflorum, spring starflower, is usually chosen by bloom color. If you are using it for naturalizing, you could choose the straight species, or a mixture of colors that may be less expensive than named cultivars. 'Alberto Castillo' is one of the largest white-flowering spring starflowers. 'Charlotte Bishop' and 'Tessa' are pink in color. 'Rolf Fiedler' has blue flowers. 'Wisley Blue' has violet-blue flowers that wash through to a pale center.

TOP: *Ipheion uniflorum* // BOTTOM: *Ipheion uniflorum*

Iris
iris

The genus *Iris* includes a wide array of good garden bulbs in a range of colors. Most of these irises have the recognizable iris shape that is vaguely triangular from above. From the side, the three center petals stand upward and are called standards. The three outer petals are often decorative and are called falls. They may have a decorative blotch or band in their center.

The bulbous irises are listed in sequential order of bloom. To begin the gardening year, there are short rock garden and Juno irises with their spring color impact. These are followed by a succession of bearded irises, including short ones that bloom first, and then on to taller ones. Dutch irises are another group of spring-into-summer bloomers that grow from a true bulb. Finally, blackberry lilies, which were in another genus, have been moved to *Iris*. The selection here includes only bulbous irises, but there are other perennial irises that are usually bought already growing in a pot.

Iris germanica 'Mary Barnett'

Iris reticulata
rock garden iris

Full sun // Well-drained soil // Zones 3–8 // Height 5–8 in. //
Season of interest—late winter through early spring

About These Bulbs

These little rock garden irises produce their flowers in late winter and early spring, alongside winter aconites and snowdrops. The plants are short, but their distinctive flowers and coloring mean that they stand out in the garden at this time of year. They are a lovely surprise to find when you are waiting for spring to fully arrive.

The most popular cultivars of *Iris reticulata* have complex flower colors of blue or deep purple, with an eye-catching pattern of white, cream, and yellow in the center of their lower petals. They have linear leaves that elongate as the flowers fade. Many of these irises are fragrant.

Planting and Maintenance

Rock garden irises require well-drained soil for good growth. Plant them in ground that is slightly raised above regular soil level, so that the tiny bulbs get a summer baking. Do not plant these irises in an irrigated flower bed. A mulch of inorganic matter like grit or gravel is preferable to wood because the stones aid water movement away from the bulbs, to stop mud from splashing onto the petals.

If you cannot provide the conditions that these irises need in the flower bed, grow them in a container of well-drained potting soil. When you plant them, add plenty of grit, gravel, or coarse sand as part of your container mix.

Plant the bulbs in clusters by digging a wide hole and spacing them out, so that when they grow, you can admire each distinct iris bloom. Put them about five inches deep, which is deeper than you might think from the bulb size. At this lower level in the soil, the bulbs stay slightly cooler than at the surface and moist but not wet. In a pot, they only need to be planted a couple of inches deep.

Design Ideas

Rock garden irises are small plants that bloom early in the gardening season, so place them where you will see them at that time of year. They look good in an elevated rock garden or gravel-mulched flower bed, which also provides them with ideal growing conditions. All of these irises need plentiful spring sunshine and a hot or warm summer to perennialize. Surrounding them with a stone mulch improves their appearance; it also reflects heat and light into their foliage, to help keep the bulbs dry. If you grow these bulbs in pots, choose

terracotta bulb pans for good drainage. When the irises are in bloom, raise them up so that you can admire the intricate flower shapes and colors, and smell their fragrance.

Species and Cultivars

Rock garden irises are mainly bred from the species *Iris reticulata*. However, there are other miniature early-flowering irises, particularly *I. histrioides*, which have been used to breed new flowers. Most of these cultivars are petite, but there are a few slightly taller ones too.

'Katharine Hodgkin' is one of the best and most distinctive of the bunch. It blooms early with flowers that are gray-blue with bright yellow middles to the lower petals. The spotted pattern adds to the interest. There is a preponderance of blue among the rock garden irises. Some of the prettiest are 'Harmony', which is mid–sky blue with darker falls and a white-and-yellow center; the slightly lighter blue 'Clairette'; and 'Alida' or 'Cantab', which are light blue with a yellow central splash. 'Pixie' stands out well against light gravel with its dark blue-purple petals and a yellow stripe. 'George' is a distinctive deep purple color, with dark falls and a white-and-yellow blotch. For white and light-colored irises, choose 'Painted Lady', 'Polar Ice', or 'Frozen Planet', which have a white base color that is speckled and washed with light blue. 'Eye Catcher' is similar and lives up to its name with a stronger blue center. It has a yellow tinge when it first emerges which gradually fades to white.

Iris danfordiae is another small iris. It has an intense acid-yellow color, spotted with mid-green. This short-statured plant blooms for me among crocuses in a never-watered, gravel-mulched spot. It is hardy in zones 5 to 9.

Juno irises are a group that blooms just after the early rock garden irises, requiring similar growing conditions of full sun and great drainage. These are not as hardy as the earliest irises, so plant them in a sheltered position if you are at the colder end of their range. Instead of narrow leaves, these irises have bold, wide leaves that are attached to the flowering stem. Their bulbs have additional fleshy roots that are quite fragile, so handle them with care when planting. *Iris aucheri* has pale blue flowers that emerge from the axils of the broad leaves. *I. bucharica* is about a foot tall and has yellow-and-cream flowers. Fragrant *I. cycloglossa*, one of the tall Juno irises, has lavender-purple blooms with yellow centers. I grow it among lavender plants in a never-watered gravel garden. Juno irises are all hardy in zones 5 to 9.

Iris ×hollandica
Dutch hybrid iris

Full sun // Moist, well-drained garden soil // Zones 6–9 //
Height 16–30 in. // Season of interest—spring

About These Bulbs

Iris ×hollandica, Dutch iris, is a hybrid of several species, and is closely related to English (*I. latifolia*) and Spanish (*I. xiphium*) irises. They are easy to grow from fall-planted bulbs. The flowers are the typical iris shape, in blue, white, yellow, purple, and dark maroon. The falls have a contrasting mark in their center that is often yellow and white. Their stems are one and a half to two feet tall and are clasped by their small leaves. They have long, strong upright stems that make them great cut flowers. They also hold up well in a vase.

Planting and Maintenance

The small bulbs are easy to plant in autumn at a depth of about four inches. If they are grown as annuals for cutting, they may be composted after bloom. If they are grown in good garden soil with drainage and full sun, they usually come back for a few years.

Design Ideas

Dutch irises are easy to grow among perennials in a mixed flower bed, or in a row in a cutting patch. If the flower and foliage are cut for arrangements, the bulb may not return next year. If you plan to cut them, you may need to buy new bulbs each year. I have had clumps that are reliably perennial for several years, and I enjoy the look of them rising through cool-season annuals, such as lady's lace (*Ammi majus*) and larkspur (*Consolida ajacis*). Dutch irises can be included in a mixed spring-flowering container to add height to the composition.

Iris ×hollandica 'Carmen' // Iris ×hollandica 'Red Ember' // Iris ×hollandica 'Blue Magic'

Species and Cultivars

Look for lovely white 'Alaska', classic 'Blue Magic', which has a yellow center to the blue petals, dark maroon 'Red Ember', bright yellow 'Golden Beauty', and purple-and-yellow 'Mystic Beauty'. One of my favorites is 'Carmen'. It has a white-purple wash, purple middle, and yellow blotch in the center of each lower petal. There are other available cultivars that are all worth growing. They have the same upright effect in the garden and vase, but they differ in their color schemes.

Iris germanica
bearded iris

Full sun // Well-drained soil // Zones 4–9 // Height varies //
Season of interest—late spring and occasionally fall

About These Bulbs

Bearded irises are the type of iris with which most people are familiar and prone to grow. Their beautiful flowers are in the classic iris shape, with fuzzy centers called "beards" on their falls. The three upright standards may be the same color or a different one. Bearded irises are available in a delightfully wide range of colors, except for true red and bright royal blue. The petals on the old-fashioned bearded irises are plain and simple in outline. The new cultivars have jazzy patterns, centers that have contrasting colors, stripes, veining, and ruffles. Some bearded irises have an intriguing soft fragrance. The flowers are held on leafless stems. The pointed, sword-shaped leaves of bearded irises are glaucous, silvery green, and are borne in a basal clump attached to the rhizome. Given the vast choice of cultivars from which to choose, you can really enjoy the selection process to find the ones that suit your style. Bearded irises are extremely hardy and deer and rabbit resistant.

Planting and Maintenance

Bearded irises are known for their drought tolerance and their need for a sunny, hot, well-drained position in the garden. They grow from their bulbs that are rhizomes. When planting, lay the rhizomes parallel to the soil surface and push a gritty soil mixture around the sides, so that you barely bury it. The soil is used to hold the rhizome in place, not to cover it. The new roots will emerge at the bottom of the rhizome. If your garden has clay or heavy loam soil, add plenty of grit or sand to the planting bed to ensure excellent drainage. Consider planting them in a raised bed because bearded irises will rot if surrounded by water-retentive soil.

Bearded irises can be divided after they have finished flowering in early summer. Trim the flower stalks and the leaves down to about six inches, to reduce the water loss from the leaves. Lever the rhizomes out of the ground carefully. Cut them into sections that each have a fan of healthy-looking leaves and some roots. Use a clean knife to cut up the rhizome and discard pieces that look rotten or moldy. Disinfect your knife between cuts with an alcohol spray. Once you have made the cuts, let them dry off for a couple

This group of bearded irises is a fantastic late spring sight. // *Iris* 'Thrice Blessed' // *Iris germanica* 'Moonlit Sea'

of hours to heal the wound before replanting. Dividing irises makes more plants for you and your friends, and it renews the vigor and flowering. If left alone, bearded irises tend to bloom at the outer ends of the rhizomes and the middle dies out. Bearded irises make great pass-along plants because they are so easy to dig, divide, and share with gardening friends.

The iris borer is a pest that wreaks havoc on bearded irises. It you see borers or there are holes in the rhizome, cut out these parts and dispose of them. If possible, replant your irises into a different part of your garden. Clean up old foliage in the autumn and inspect the rhizomes at the same time for rotted areas.

Design Ideas

Tall bearded irises are the most commonly grown, but there are also midsize and even short ones. Plant a variety of sizes to achieve a succession of bloom. Short bearded irises usually bloom first and are self-supporting. They look great in a sunny gravel garden or a raised rock garden.

When they bloom, tall bearded irises are beautiful, statuesque showstoppers. Their growing needs are very specific, so they are often grown in a bed with other bearded irises. I grow some in a raised mixed bed that contains alliums. These two plants bloom at the same time and look good together. Another approach is to integrate them into a dry garden with other drought-tolerant plants. The flowering stems of tall irises can be two or three feet tall so they may need staking or tying to a fence.

OPPOSITE: *Iris germanica* 'Moon Sparkle'

Iris germanica 'Shasta' // *Iris germanica* 'Cee Jay' // *Iris germanica* 'Mary Barnett'

Species and Cultivars

Bearded irises are available in an extensive number of cultivars. The ones listed here are just a few of the thousands that are available. Check the plant height, flower colors, shapes, and scents when buying. If possible, go see them in person and buy when they are in flower.

Here is a selection of the plethora of bearded irises from which you can choose. 'Before the Storm' has very dark flowers that are almost black. At the light end of the spectrum, choose the white 'Shasta', or the reblooming 'Immortality', which has a second flush of flowers in the autumn. Other reblooming irises include 'Cantina', 'Orange Harvest', and 'Buckwheat'. One of the most popular bearded irises is 'Beverly Sills', with coral-pink flowers. For a brilliant orange iris, try 'Octoberfest'. There are plenty of irises in the purple and blue range, including 'Jane Phillips', a popular mid to pale blue-purple, and the deep dark purple 'Titan's Glory'. 'Cee Jay' is a spectacular purple and white, and a vigorous grower. For yellow, you could try 'That's All Folks', with its large yellow-gold flowers, or the bicolored yellow-and-white 'Moon Sparkle'. In a rock garden, I grow the short early iris 'Thrice Blessed', which has yellow flowers with pale blue beards on the falls.

There are irises that have falls that are one color and standards that are another. One example is 'Edith Wolford', which has light yellow standards and mid-purple falls. Others have veining patterns in contrasting colors, like those seen on the scented, unusually colored maroon and peach 'Artistic Web'.

For those of us who love historic plants, look for bearded irises that were bred decades ago, which still make super garden plants. A few

OPPOSITE: *Iris germanica* 'Oktoberfest'

Iris germanica 'Before the Storm' // *Iris germanica* 'Artistic Web'

examples include the soft lavender-and-yellow 'Mary Barnett', from 1926; 'Louvois', which is a rich brown bicolor, bred in 1939; or the striking purple-and-yellow 'Moonlit Sea', from 1942. Whatever iris you choose, you will look forward to seeing it bloom each spring.

Iris domestica

belamcanda, blackberry lily

Full sun // Well-drained soil // Zones 5–10 //
Height 24–36 in. // Season of interest—summer

About These Bulbs

These plants are now included in *Iris* but were formerly in the genus *Belamcanda*. They are grown for their summer-blooming six-petalled flowers, which are orange or sometimes yellow with darker speckles. The flowers do not look like the other bulbous irises. They open during the day and close at night. The pollinated flower screws itself up like a tiny corkscrew that drops off to leave the growing green ovary below. This swells and gets larger until it bursts open to reveal the shiny black seeds, which give this plant the common name of blackberry lily. These inedible seeds are arranged around the top of the stem just like the berry. The plants have sword-shaped foliage that is bright green.

Planting and Maintenance

Blackberry lilies are normally bought as potted plants, but once you have them established in your garden, you can make many more. I have had the most success growing them when I sprinkle the individual black seeds into the garden immediately after they ripen. You might see seeds rolling around beneath the plant. The best plants grow from seeds that fall between paving cracks or in a layer of gravel mulch over soil. They need a place in full sun with good drainage to flower well.

Design Ideas

These short-lived perennials are most at home when grown in dry soil that contains some rock or gravel. They often grow singly, rather than as a big clump, so intersperse them between summer-blooming perennials and annuals. Their fan-like leaves are a decorative feature.

Iris domestica // 'Hello Yellow'

Species and Cultivars

The flowers are usually orange, but 'Hello Yellow' has yellow flowers that are not spotted, held on a shorter plant than the species.

Leucojum

leucojum, snowflake

Part sun // Moist to average soil // Zones 4–9 // Height 8–20 in. // Season of interest—late winter through spring

About These Bulbs

Snowflakes are useful spring bulbs for a partly shaded area with moist soil. They have cute, downward-facing, bell-shaped white flowers with small green dots on the lower edges. Two species are generally grown in gardens: the early-blooming spring snowflake (*Leucojum vernum*), and the later-flowering summer snowflake (*L. aestivum*), which actually blooms later in spring.

Planting and Maintenance

Plant snowflake bulbs in autumn, and they will bloom the following spring. The bulbs tend to dry out, so plant them as soon as they arrive, in a flower bed that contains moist soil. To increase the moisture-holding capacity of the soil, add organic matter like leaf mold or compost to the planting bed. Once the weather begins to warm up, spring snowflakes emerge quickly. Remove wet leaf litter that might be covering the emerging tips of the snowflakes.

Leucojum bulbs grow vigorously to make large clumps that can be dug and divided after a few years. To propagate your existing bulbs or to share with a friend, dig up the whole clump and pull it apart into sections with your fingers. It is best to leave a few bulbs clinging together because they seem to establish faster that way. Replant each grouping of bulbs into moist, organic-rich soil as quickly as possible.

Design Ideas

Snowflakes are used for naturalizing in a partially shaded situation, with moist soil. They look good when planted in clumps around a pond or under a downspout, or when combined with other spring bulbs along a woodland edge. In warm winter climates, they are a good alternative to snowdrops. I grow them in a pond-side bed with checkered fritillaries and primroses (*Primula*).

Species and Cultivars

Leucojum vernum, spring snowflake, is the earliest to bloom. This is the shorter of the two snowflakes, growing to about eight inches. It blooms at about the same time as snowdrops. Spring snowflakes form a perfectly symmetrical white bell with six pointed petals that dangle from a flexible green stem. At the bottom of each point is a tiny green or yellow mark. There can be one or two of these pristine flowerheads per stem.

Leucojum aestivum, summer snowflake, actually flowers in spring, after *L. vernum*. This tall plant grows to 20 inches or more, especially in the case of the cultivar 'Gravetye Giant'. The term "giant" may be rather a misnomer, but it is a vigorous old cultivar that was named for Gravetye Manor in Sussex, England, where garden author and well-known horticulturist William Robinson lived and gardened. This snowflake has escaped from cultivation in some areas of the world, so check your local invasive plant lists before purchasing.

Muscari

muscari, grape hyacinth

Full sun to part sun // Moist, well-drained soil // Zones 4–9 //
Height 6–12 in. // Season of interest—spring

About These Bulbs

Muscari are short spring flowers that are easy to grow. They usually have cobalt blue flowers, but some come in other colors including soft blue, light pink, white, and one that is purple and yellow.

From a distance, the overall shape of each flower is a narrow upright column, but as you look closer, there are individual, often urn-shaped, flowers that are clustered in an attractive pattern up the stem, somewhat like a bunch of grapes. Grape hyacinths make a lovely little cut flower. They combine nicely with miniature daffodils and other small spring bulbs, in a vase or in the garden. Early flying pollinators are attracted to these flowers for nectar. Some muscari flowers have a honey or musk-like fragrance.

Planting and Maintenance

Grape hyacinths are easy bulbs to grow in average garden soil and spring sunshine. The bulbs are planted in autumn about four to five inches deep. While the plant is in flower, the long, narrow, bright green leaves are low to the ground. They then wither away for the summer but may reemerge during fall or winter. Muscari bulbs can be dug up by rodents, so try putting them in the same planting hole as daffodils for protection.

Design Ideas

Grape hyacinths make good companion plants for other spring-blooming bulbs. They are traditionally planted among daffodils or tulips. Use them in repeated groups throughout a flower bed or as a path edging. We combine grape hyacinths with blanda anemones, ipheions, and crocuses to give a long season of bloom at the front of a mixed flower garden.

Grape hyacinths are easy to grow in containers, either with one cultivar per pot, or planted around taller bulbs, to provide a contrast in color and flower shape. When planting up containers and window boxes, use muscari around the edges so that the foliage can dangle over the rim.

CLOCKWISE FROM LEFT: Grape hyacinths look great when planted in groups. // *Muscari armeniacum* 'Julia' // *Muscari armeniacum* 'Marleen' // *Muscari armeniacum* 'Touch of Snow'

Species and Cultivars

Muscari armeniacum is the most commonly available species. It has fragrant blue flowers that are gathered at the top of the stems. A strong grower, it is taller than some other grape hyacinths, growing to about eight inches. It can be naturalized on the sunny side of azaleas and other shrubs or planted in weak lawn areas that get full sun in spring. Watch this prolific seeder for invasiveness. Some attractive cultivars include 'Julia' and 'Touch of Snow', which are both blue at the bottom and have a greenish-white top that gets whiter with age. Each urn-shaped individual flower is rimmed in white. 'Marleen' has cobalt blue flowers with tiny touches of white at the flower edges. 'Siberian Tiger' is all white, and the delightful 'Valerie Finnis' is a soft powder blue.

'Blue Spike' is a double-flowered cultivar with densely packed heads.

Muscari aucheri produces short, attractive flowers. There are some lovely cultivars including 'Blue Magic', which is, of course, blue, and 'Ocean Magic', which is pale blue and white. The flowers of 'White Magic' emerge from the ground as light green buds, opening to white.

Muscari latifolium is a distinctive plant that has one large, wide leaf that grows up beside the flowers. As it emerges from the ground, the top of the leaf acts like a hood that shelters the emerging flower. The flowers are notable because they are two-toned, with a topknot of mid-blue sterile flowers above dark blue-violet flowers. I love this one in pots, and in the garden combined

CLOCKWISE FROM LEFT: *Muscari aucheri* 'White Magic' // *Muscari latifolium* // *Muscari* 'Alaska' // *Muscari azureum*

with tall daffodils like 'Cragford'. It does best in part sun with some summer moisture. The cultivar 'Grape Ice' has unusual deep purple lower flowers and a sprig of green on top that opens to white.

Muscari azureum has a different look. Its flowers flare open at the ends and are sky blue, with a dark blue stripe. We love to grow this one in containers, so that you can see the intricate details.

Muscari botryoides is another commonly grown grape hyacinth. Each flower looks like a tiny urn. The flowering spikes become looser and elongate as the flower grows. Try the cultivars 'Album' and 'Superstar'.

Muscari comosum 'Plumosum' is unlike other grape hyacinths because its flowers are not bell-shaped but look fluffy. The flower color is violet-purple. *M. macrocarpum* is another oddity that opens purple and matures to bright yellow. It needs to be grown in a sunny rock garden. Do not irrigate it in summer because the bulb needs a period of baking to do well. It is one of the most fragrant of the grape hyacinths. It is less hardy than others in this group. The cultivar 'Golden Fragrance' is often available.

Other useful muscari cultivars to look for include the hybrid 'Pink Sunrise', which has delicate pastel pink flowers; 'Baby's Breath', with light blue flowers that emerge green; and 'Alaska', which is white with a wash of pale blue.

Narcissus

daffodil

Full sun to part sun // Moist, well-drained soil // Zones 3–9//
Height 4–24 in. // Season of interest—spring

About These Bulbs

Daffodil time is a highlight of the bulb gardening year. Daffodils are carefree plants with cheerful blooms. The general perception is that daffodils are yellow and have a trumpet shape. Gardeners who do not grow many daffodils are surprised at the available diversity of height, form, flower shape, and color. Daffodil size ranges from small to tall. Colors vary from white to yellow, with some added orange, green, red, or pink accents. Some daffodils have one head to a stem, while others have multiple heads. The best way to determine what you like is to visit a garden that grows daffodils, so you can see them when they are in flower. I might go so far as saying that every garden needs at least one patch of daffodils.

With careful selection, you can have daffodils in bloom for months. To get a long season of bloom, plant multiple daffodil types. If temperatures remain cool, each individual daffodil bloom may last for a long time. Buy some cultivars that bloom in early spring, others for midseason, and some late ones. I have different daffodils in bloom in my garden for about four months. They also make great cut flowers. If you choose daffodils that are suited to your climate, they will survive for years. I like to grow old-fashioned, or historic, daffodils, which have stood the test of time. A historic daffodil is defined as one that was bred and introduced before 1940. Decades after their introductions, they continue to make excellent garden plants. One reason daffodils do well in gardens is pests do not eat them. Daffodils produce a chemical that makes them unpalatable to most burrowing and browsing mammals.

Planting and Maintenance

Plant daffodils in autumn in holes that are three times the height of the bulbs, in average- to free-draining soil, in full sun. The width of the hole will vary according to how many bulbs you want to plant together. Some people dig one hole per bulb, but I like to make small groups of three or five and plant them in the same hole. Small ones should be planted closer together than large ones. After the bulbs have finished flowering in late spring, leave the foliage in place to nourish them for next year. You can deadhead the flowers if you wish. Ensure that the planting area stays moist, but not wet, in the summer. There are a few species that can grow in damp soils.

Design Ideas

Daffodils are versatile plants for the spring garden, both in the ground and in containers. They can be used in bedding schemes as annuals or left in the ground as perennials. They range in size from tiny ones that are classified as miniatures to the bold and bright ones that are great for a mass planting. Miniature or small ones look good at the front of a bed, in a rock garden, or around the edge of a large pot. Tall daffodils are best planted in the middle of a mixed planting. You can see the large daffodil blooms because they come up earlier than surrounding plants. Later on, the emerging perennials disguise the fading daffodil foliage. Choosing a few types in your chosen color scheme is often more successful than a jolly mixture. If you create a daffodil lawn, use ones with a delicate wildflower look, rather than those that have heavy heads or are garish. Daffodils are so lovely against the fresh green of spring grass. My youngest daughter dreams of one day having a daffodil meadow, after walking through the ones at the public gardens at Winterthur in Delaware and Filoli in California.

Species and Cultivars

Daffodils are divided into the following groups, called divisions, that help with identification. Knowing the details of the divisions is not necessary to grow and enjoy beautiful daffodils, but they are a useful aid when choosing your daffodils. As you read the descriptions, you will see that the flowers are divided into these groupings using a variety of factors such as the length of the trumpet, the number and size of the flowers on each stem, the colors, fragrance, and time of bloom.

Daffodil flowers range in color from white through cream, light yellow, deep yellow, and orange. Their cups may additionally be any of those and have coral-pink, green, or red, especially as a rim. Colored cups emerge with bright fresh colors and then fade in strong sunshine. Fragrances vary from the classic daffodil scent that just smells like spring, to the slightly sweet smell of the Jonquilla Group, or the scent-pumping flowers of some tazettas.

Division 1, trumpet daffodils, are the classic daffodils that have a long trumpet-like middle cup that protrudes beyond the rest of the flower. Each flower is held on its own stem. They are early bloomers in the garden. There is a lot of yellow in this division, with some of them being so brightly colored that it is hard to distinguish the cup shape from the surrounding petals. Flower sizes range from large and bold to miniatures that have the same proportions and the same long trumpet. Stem length for large trumpet daffodils is 18 to 20 inches tall; small ones, such as 'Little Gem', reach about 6 inches.

The earliest daffodil to bloom in my garden is in this group and is called 'Rijnveld's Early Sensation'. It blooms alongside snowdrops and crocuses. 'King Alfred' is an old-fashioned favorite that set the standard for daffodils from this group. Today, it is often replaced by similar daffodils like 'Dutch Master', which have similar yellow flowers and a bold presence in the garden. A good white trumpet daffodil is 'Mount Hood', which opens slightly cream and matures to white. For an unusual color choice, try 'Fidelity' with its light yellow outer petals and a pinkish peach

trumpet. I love the historic yellow-and-white 'Empress', which dates from 1869.

Division 2, large-cup daffodils bloom slightly later than the trumpet types. The cups of these daffodils are more than one third the length of the outer petals, but are less than equal to them. To test this, bend the outer petals around the cup, and look at their relative lengths. The cup should not protrude beyond the outer petals. There is only one flower on each 16- to 20-inch stem. This is a popular division in gardens and containers with many good cultivars available.

One of the most popular in this division is 'Ice Follies', which has soft cream outer petals and a yellow cup that fades to cream. If you want strong colors, try those with bright yellow outer petals and an orange cup like 'Ceylon', 'Fortune', or 'Brackenhurst'. For white daffodils in this division, look at 'Stainless' and 'Snowboard'. 'Flower Record' is cream with a yellow cup and an orange rim. Some daffodils in this division have what is called a "pink" cup, which, to my eye, looks coral. 'Pensioner' is one such example. This type should be planted in part shade for longest-lasting color. If you love scented daffodils, choose 'Fragrant Rose'. A good historic daffodil is 'Croesus'. I grow it in a raised, gravel-topped bed.

Division 3, small-cup daffodils, have a cup that is less than a third of the length of the outer petals. The overall look of the flowers in this group is graceful and charming. The flowers are borne one to a 14- to 16-inch stem. The flowerheads are lightweight and tend to stay upright, even after a spring storm. Daffodils in this division usually have creamy white or yellow outer petals. Their

CLOCKWISE FROM LEFT: *Narcissus* 'Conspicuus' // *Narcissus* 'Xit' // *Narcissus* 'Flower Drift' // *Narcissus* 'Tête Bouclé'

cups range from small to tiny in size and are available in a variety of colors.

'Barrett Browning' is a good grower with white outer petals and a small orange cup. 'Goose Green' and 'Dreamlight' are both white; their cups are rimmed with red, shading to yellow and green in the center. For yellow outer petals and an orange cup, grow 'Conspicuus'. 'Polar Ice' is a nearly white daffodil with a faintly yellow cup. I love the old daffodil 'White Lady', which has a small yellow cup, and 'Segovia', which also has a yellow cup on a short stem. The cutest, in my opinion, is the tiny all-white 'Xit', which stands at about six to eight inches.

Division 4, double daffodils, have a full fluffy look with lots of extra petals. Some have one flower to a stem, while others have multiple heads. Stem lengths can vary from 8 to 24 inches.

The extra petals can make their heads heavy, so daffodils in this division should be planted in a sheltered position, so that they are not blown over by strong winds.

One of the most popular in this group is the fully double 'Tahiti', which has bright yellow petals with smaller orange petals tucked between them. They are good daffodils to use where you want a bold effect from a distance. Choose 'Flower Drift' if you want a white double with lots of little orange petals in the middle. Some double daffodils have many petals inside the trumpet, such as 'Petit Four', with white outer petals and a soft coral center. Doubles with multiple heads include the scented white-and-orange 'Bridal Crown'; 'Cheerfulness', which has creamy yellow petals and little yellow-orange petals at the center; and 'Yellow Cheerfulness', which is similar to the former, with soft yellow petals. The short yellow

CLOCKWISE FROM LEFT: *Narcissus* 'Hawera' // *Narcissus* 'Lemon Drops' // *Narcissus* 'Emcys' // *Narcissus* 'Orange Comet'

daffodil 'Tête Bouclé' is useful in pots or at the front of a border.

Division 5, triandrus daffodils, have two or more nodding heads per stem and multiple stems from each bulb. These daffodils give very good value, producing a floriferous show either in the garden or in pots. Each little flower has a wide but short cup that is surrounded by slightly turned-back, graceful outer petals. The flowers often have a delightful fragrance. The foliage is slender and dies down readily after bloom. These characteristics mean that triandrus daffodils are one of the top candidates for easy integration into a mixed flower bed. They are also useful because they will grow in part shade. Plants in this division vary in height from 8 to 16 inches.

'Thalia' is one of the most popular triandrus daffodils. It is all white and increases successfully in the garden. It is a historic daffodil that dates from 1916. Some newer white daffodils in this group include 'Starlight Sensation' and 'Elvin's Voice'. I have pale yellow 'Lemon Drops' growing around redbud trees (*Cercis canadensis*). They bloom at the same time and look lovely together. A personal favorite is the slender, delicate, light yellow 'Hawera'. Use this in a rock garden area or as a container plant.

Division 6, cyclamineus daffodils, have swept-back-outer petals that look perky and full of personality. They bloom early in the daffodil season on 6- to 20-inch stems. Use these daffodils in containers or in raised beds, such as rock gardens.

This is one of my favorite divisions. The flowers have such personality and a sense of forward motion. When I make up my annual list of daffodils to buy, I always include some

CLOCKWISE FROM TOP LEFT: *Narcissus* 'Kokopelli' // *Narcissus* 'Trevithian' // *Narcissus* 'Avalanche' // *Narcissus* 'Oxford Gold' // *Narcissus* 'Geranium' // *Narcissus* 'Ornatus' // *Narcissus poeticus* var. *recurvus*

cyclamineus. One of the first to bloom is the all-yellow 'February Gold'. 'Rapture' is known for its early egg-yolk-yellow blooms and its ability to multiply in the garden. There are some great colors in this division, such as 'Jetfire', which has yellow outer petals and an orange cup, and 'Orange Comet', which also has an orange cup surrounded by creamy white outer petals. For a subtle coloration, try the creamish white 'Emcys', with a cup that is light yellow fading to white. 'Surfside' has a similar color but is a useful short daffodil for pots or the front of a border.

Division 7, jonquilla daffodils, are sweetly scented and usually have between two and six flowers per stem. The cups are small and may be tilted slightly down. Jonquils are beloved in southeastern American gardens where they grow well and naturalize readily.

Some of the most popular jonquils are the bright yellow cultivars 'Sweetness' and 'Quail'. These have green, narrow, rush-like foliage that gets tucked away easily in a mixed border after flowering has finished. The cultivar 'Kokopelli' is also yellow but with a darker yellow-orange cup. The plant looks full of flowers because each bulb produces several stems, each bearing four flowers. 'Pipit' is another favored jonquil that has white cups that feather into the yellow outer petals. 'Golden Echo' has a yellow cup that washes into a creamy set of outer petals. Some jonquils bloom late in the season like 'Sun Disc', which has small flat faces, and 'Baby Moon', which has rich golden yellow flowers. A strong-growing historic that I love is 'Trevithian', which dates from 1927.

CLOCKWISE FROM LEFT: *Narcissus 'White Petticoat' // Narcissus 'Mondragon' // Narcissus 'Mary Gay Lirette' // Narcissus 'Tête-à-Tête'*

Division 8, tazetta daffodils, have multiple heads per stem and a characteristically strong smell. Tazetta daffodils tend to be less hardy than other divisions but are well suited for outdoor growth in climates with mild winters. 'Avalanche', 'Geranium', and 'Cragford' are some of the hardiest, and can be grown in zones 5 or 6 to 9. For inside blooms, choose the tender paperwhites.

Division 9, poeticus daffodils, are some of the latest to bloom. They show up well in gardens because they have outer petals that are relatively large, flat-faced, and crystalline white. Their button-sized central cups are cute, with concentric rings of red, green, and yellow. Poeticus daffodils are a great choice if you want to naturalize daffodils in lawns. The best one for this purpose is the old-fashioned *Narcissus poeticus* var. *recurvus* (pheasant's eye daffodil). Its petite flowerheads with swept-back and slightly twisted outer petals are graceful among grasses. These daffodils also look great in mixed flower beds. 'Actaea', 'Cantabile', and 'Horace' are some of the best. For an early-blooming poeticus, choose the historic 'Ornatus'. It is tolerant of both cold and warm winters.

Division 10, bulbocodium daffodils, barely look like daffodils. Their flowers are all cup with tiny vestigial strands for the outer petals. They have short stems that are only about six inches high. There is one flower per stem. Try this daffodil in small groups on the sunny side of a group of shrubs, at the front of a border, or in moist grass. The cultivar 'Oxford Gold' seems to be particularly adaptable to garden culture, but I also like 'White Petticoat'.

Narcissus 'Tiny Bubbles' // Narcissus ×medioluteus // Narcissus romieuxii

Division 11, split-corona daffodils, are something different. This division can be divided into two subdivisions. The first group, (a), has a rather flat-faced flower, with a full center. This is called the collar type. I have been growing 'Mary Gay Lirette' for years; recently I have also been taken with the pale greenish yellow 'Exotic Mystery' and the fragrant all-yellow 'Tripartite'. If you want a bright split-corona daffodil, choose the yellow-and-orange 'Mondragon'. For a more subtle peach-and-cream color scheme, pick 'Apricot Whirl'.

The second group within this division, (b), is called the butterfly or papillon type. These flowers look like they have a little star in the center. The tips of the inner petals line up with the gaps between the outer petals. One of my favorites is 'Lemon Beauty', with white outer petals and cream-and-yellow inner markings.

Division 12, miscellaneous daffodils, are those that do not fit exactly into other divisions. 'Tête-à-Tête' is one of the most popular daffodils in the world and is widely used as a container plant. 'Tiny Bubbles' is a miniature that is wonderful in a raised bed. 'Bittern' is a strong plant with multiheaded, yellow-and-orange flowers.

Division 13, species daffodils, are the original wildflowers that were brought into gardens. They have been used for breeding to develop all of the garden cultivars that we have today. *Narcissus romieuxii* is a lovely pale yellow flower with a light citrusy scent. *N. ×medioluteus* has two flowers per stem and is bright yellow. *N. pseudonarcissus* is a wild white-and-yellow trumpet-style.

Ornithogalum
ornithogalum, star of Bethlehem

Full sun to part shade // Rich, well-drained soil // Zones 5–10, varies // Height 8–36 in. // Season of interest—spring

About These Bulbs

Ornithogalum are known for their star-shaped flowers which are white to cream with green accents in most species. There are some that are orange or yellow as well. Most of the flowers close at night, and then open again when the sun comes out. Their heights and overall looks vary as does their hardiness. Add tender species to the garden in pots that you can bring in for winter, or use them as a temporary occupant of the flower bed. Hardy species make great plants for naturalizing in a wild area of the garden. However, some species can be invasive, so check your local listings before planting.

Planting and Maintenance

If the *Ornithogalum* species that you have chosen is hardy where you garden, plant them outside in the autumn, in flower beds or in pots. If they are not hardy in your zone, wait until spring and add them as tender plants to your containers or beds. Dig them up again in autumn to save inside until the following spring.

Design Ideas

Tall species can be used as accents, repeated throughout a bed, combined with lower-growing bulbs. Short ones are great in containers. Some of them make good cut flowers, so they can be used in a cutting garden.

Species and Cultivars

Ornithogalum ponticum is an airy plant that has starry white flowers borne in conical sprays. Its airiness allows it to grow through surrounding plants, which makes it a great addition to a flower bed. It is about three feet tall, flowering in late spring. I grow the cultivar 'Sochi' in a mixed flower bed, where it grows through betony (*Betonica officinalis*) and rose campion (*Silene coronaria*).

Ornithogalum ponticum 'Sochi' // *Ornithogalum dubium* // *Ornithogalum saundersiae*

Ornithogalum dubium looks unlike the others in this genus because it is bright orange or yellow. It is hardy in zones 8 to 10, so many gardeners grow it in containers for winter bloom inside. The whole flower is larger than other members of this genus and has a bold presence.

Ornithogalum saundersiae, giant chincherinchee, with its fabulously musical-sounding common name, has a white hemispheric flowerhead composed of many florets, each with a dark green center. It blooms in summer and reaches three to four feet tall. It is hardy in zones 7 to 10. Bring it inside for winter if it's not hardy in your garden.

Ornithogalum umbellatum, star of Bethlehem, is native to parts of Europe and North Africa where it can be grown in naturalized swaths. Avoid planting in North America, where it is extremely invasive. It grows to about six or more inches tall, with a characteristic silver stripe running down the leaves and white flowers. *O. nutans*, drooping star of Bethlehem, has stems that are about a foot tall, each bearing about 20 bell-shaped flowers. The blooms are green on the outside with a creamy white interior. Wild-looking patches of this bulb can sometimes be found in old gardens.

Ranunculus asiaticus

ranunculus, Persian buttercup

Full sun // Rich, evenly moist soil // Zones 8–11 //
Height 12–18 in. // Season of interest—spring

About These Bulbs

Ranunculus asiaticus, Persian buttercup, is a showy bulbous member of the big buttercup genus, which are often found growing in fields and wild places. These buttercups are the opposite of a wildflower. They have been bred to have vibrant, rounded flowers in a variety of colors like rich purple, white, red, pink, yellow, orange, and bicolors, with some that are edged in a contrasting color. This ranunculus comes from areas where winters are not too cold, nor summers too hot. It makes a lovely addition to a cool-season planting, blooming in mid-spring, and as a beautiful cut flower. The exact flowering time will depend on when you start the bulbs into active growth.

Planting and Maintenance

In climates with mild winters, the bulbs can be planted in the autumn. In places with cold winters, they must be started in spring. Their bulbs look like a bunch of little fingers. Soak them for a few hours in tepid water until they plump up. Then, plant them about two inches deep and four to six inches apart in well-draining moist soil with full sun. They will take three or four months to come into bloom. Keep the soil moist, but not wet, so that the bulbs do not rot.

Persian buttercups need regular deadheading to promote the production of new buds. The most important thing is to water them thoroughly if the soil gets dry. Add some low-nitrogen fertilizer to encourage the growth of flowers, rather than leaves. When temperatures rise in the summer, they will stop blooming.

Design Ideas

Ranunculus asiaticus is best treated as an annual for your spring bedding scheme, that is discarded after it blooms. It makes an excellent container plant and cut flower. Combine it with other bulbs, such as miniature daffodils and poppy anemone, or cool-season annuals such as English daisies (*Bellis perennis*) and pansies (*Viola*). It goes well with herbaceous plants like primroses or polyanthus (*Primula*), snapdragons (*Antirrhinum*), wallflowers (*Erysimum*), and alumroot (*Heuchera*). Add some of these lovely bulbs to a sunny part of your vegetable area and use the cut flowers in arrangements.

A bronze-flowered *Ranunculus asiaticus* // Red and pink *Ranunculus asiaticus*

Species and Cultivars

Ranunculus asiaticus, Persian buttercup, is available in many different rich and pastel flower colors, either in single colors or in mixes. I like to select bright pinks and reds, which are not readily found in other spring flowers, to bring some bold colors into my garden and vases. You can choose fun colors that suit your displays or arrangements. The bold rounded shape is a lovely contrast to other flowers that bloom at the same time of year.

Scilla

squill, scilla

Full sun to part shade // Moist, well-drained soil // Zones 3–8 // Height 4–10 in. // Season of interest—spring

About These Bulbs

Scillas and similar bulbs are mainstays of the early spring garden. They are great choices to grow around taller bulbs. These petite beauties are inexpensive to buy, easy to plant, and cold hardy. Their bell- or star-shaped flowers are composed of six petals. The flowers are clustered along a short stem, with strap-like leaves at ground level. Blue is the most popular color for these little bulbs, but there are others in light pink, violet-purple, and pure white.

Scillas look good when planted in masses in a woodland area, at the edge of a shrub border, at the front of a flower bed, or in a container. The genus *Scilla* now includes flowers that used to be called *Chionodoxa*. The similar bulbs of *Puschkinia* are also included here because, to a gardener, they look very similar and fill the same roles.

Planting and Maintenance

Plant these little bulbs in well-drained soil in the autumn. Look at the height of the bulb and bury it in a hole three times that depth, usually about three inches. Cluster bulbs into small or large groups, rather than planting each one separately, to make an impact. Ensure that the soil has plenty of compost, leaf mold, or other organic material. Choose a location where these little bulbs can seed into the surrounding area over the next few years. These plants are maintenance-free, especially if planted between later-blooming perennials that disguise their leaves as they die down.

Design Ideas

This group of small spring bulbs is fantastic for naturalizing in open woodland. A sea of these blue flowers in the dappled light of an old woodland, shaded by deciduous trees, is a sight not to be missed. Use scillas at the front or mid-front of a bed, scattered between taller spring bulbs like hyacinths or daffodils. They look especially lovely along a path, or clustered next to a bench. It is best to plant scillas in clumps or drifts to increase their visual impact.

Scillas are wonderful in containers, with one type per pot, or planted together with daffodils or tulips. After flowering, take them out and plant them in the ground, or let the foliage die down, keeping the whole pot in an out-of-the-way place until next year.

CLOCKWISE FROM LEFT: *Scilla bifolia* // *Scilla bifolia* 'Rosea' // *Scilla forbesii* 'Violet Beauty' // *Scilla luciliae* 'Alba' // *Scilla mischtschenkoana*

Species and Cultivars

Scilla bifolia, twin-leaf squill, is a personal favorite of mine. The early-blooming inflorescence, which is pointed at the tip and wider at the base, is composed of individual six-petalled starry flowers that open in sequence from the bottom to the top. The plant has two strap-shaped leaves. The blue is delightful, but I also love the light pink, slightly scented 'Rosea'.

Scilla forbesii, glory-of-the-snow, was formerly known as *Chionodoxa* and may still be listed by this name. The plant has star-like flowers with a contrasting white center. Glory-of-the-snow lives up to its name, blooming early and welcoming pollinators to their attractive flowers. My favorites are 'Violet Beauty', which is a lovely lilac-purple with a white eye, and the white 'Alba'. We also grow the slightly larger cultivars 'Pink Giant' and 'Blue Giant'. Some similar plants may be listed as *S. luciliae*. Buy them by color and flower shape rather than by name, which seem to vary from seller to seller.

Scilla mischtschenkoana, Misczenko squill, is an early bulb that starts flowering while it is still emerging from the soil. It has pale silvery blue flowers with a central darker blue line running the length of the petal. The stem and leaves elongate after flowering. Place this short plant toward the front of a flower bed, or in a rock garden.

Scilla sardensis, lesser glory-of-the-snow, has bright blue flowers with tiny white centers that help make it visible from a distance. The flowers are clustered along the attractive dark purple-brown stems. We have this little beauty

CLOCKWISE FROM LEFT: *Scilla sardensis // Scilla siberica 'Spring Beauty' // Scilla siberica 'Alba' // Scilla peruviana*

planted beneath witch hazels (*Hamamelis*), with snowdrops nearby.

Scilla siberica, Siberian squill, is the most commonly grown garden plant of this group. It is a wonderful naturalizer that grows well in part shade, under deciduous trees, or in thin lawns, where it will seed itself in to make soft carpets of blue flowers. A choice cultivar is the beautiful blue 'Spring Beauty'. It has one of the clearest blues available in the early spring garden. Siberian squills are hardy down to zones 2 to 3, so they are favorites of cold-weather gardeners. The white cultivar, 'Alba,' is another good choice but is slightly less prolific. Each bulb produces several flowering stems.

Scilla peruviana, Peruvian squill, is a tender bulb that blooms in late spring and is hardy in zones 8 to 10. The bulbs and the plants are larger than the early-blooming squills. The flowers are purply blue, conical in shape, and borne on stems that are over a foot tall. They have real presence in the garden because the individual starry flowers gradually open in sequence from the bottom upward. Peruvian squills need full sun in a rock garden or equivalent site to bloom well. If they have sufficient moisture, they will remain evergreen. A white cultivar is available.

Puschkinia scilloides and **P. scilloides var. libanotica**, striped squill, have very pale blue flowers with a dark blue line down the middle. Their low-growing leaves are held in pairs. These bulbs are not in the genus *Scilla* but are similar in appearance and stature. They need the same growing conditions as the short scillas and look extremely similar to them. They multiply quickly when conditions are good for growth. We grow them on a raised bank so that the flowers are clearly visible. The cultivar 'Alba' has white flowers and shows up well in part shade. They look fabulous when planted with white daffodils. I also love them in containers and in bulb lawn plantings. They are hardy in zones 4 to 8 and are about four to six inches tall.

OPPOSITE: *Puschkinia scilloides* var. *libanotica* 'Alba'

Triteleia

triteleia, brodiaea

Full sun or part sun // Rich, well-drained soil // Zones 6–10 //
Height 12–28 in. // Season of interest—late spring

About These Bulbs

Triteleias grow naturally in grasslands in western North America. They are a fantastic addition to gardens in this geographic area, but also take to cultivation in other places. Provide them with spring moisture (while they are in active growth) and keep them dry in summer (while they are dormant). They bloom on leafless stems that hold a cluster of star-shaped flowers in colors ranging from blue through blue-purple and white. They used to be classified as *Brodiaea* and you may still see them for sale under this name.

Planting and Maintenance

Plant the bulbs in the autumn about three to five inches deep in well-drained soil. Add extra grit and gravel to the planting mix to ensure that the bulbs stay relatively dry during their summer dormant period. Dig a hole and put a few bulbs in the same hole spaced an inch or two apart.

Design Ideas

Triteleias are wonderful flowers for naturalizing in a well-drained rock garden, raised beds, or along the front of a mixed flower bed. The slender blooms look good when clustered together, or grown among perennials like penstemon (*Penstemon*), and the low-growing sun-loving dianthuses (*Dianthus*). I grow them in a well-drained area that is mulched with gravel. Another option is to grow these bulbs in pots with a very gritty soil. Grow them alone in a pot, or place them at the center of a mixed bulb container, where the surrounding spring bulbs will disguise the deteriorating triteleia leaves that begin to die down while the bulbs are in flower.

Species and Cultivars

Triteleia laxa is the species that is most often grown as a garden plant. Cultivars to look for are the purply blue 'Queen Fabiola', the slightly bluer 'Corrina', and the white 'Silver Queen'. They all have the same habit and look in the garden apart from flower color.

TOP: *Triteleia laxa* 'Corrina'
BOTTOM: *Triteleia laxa* 'Silver Queen'

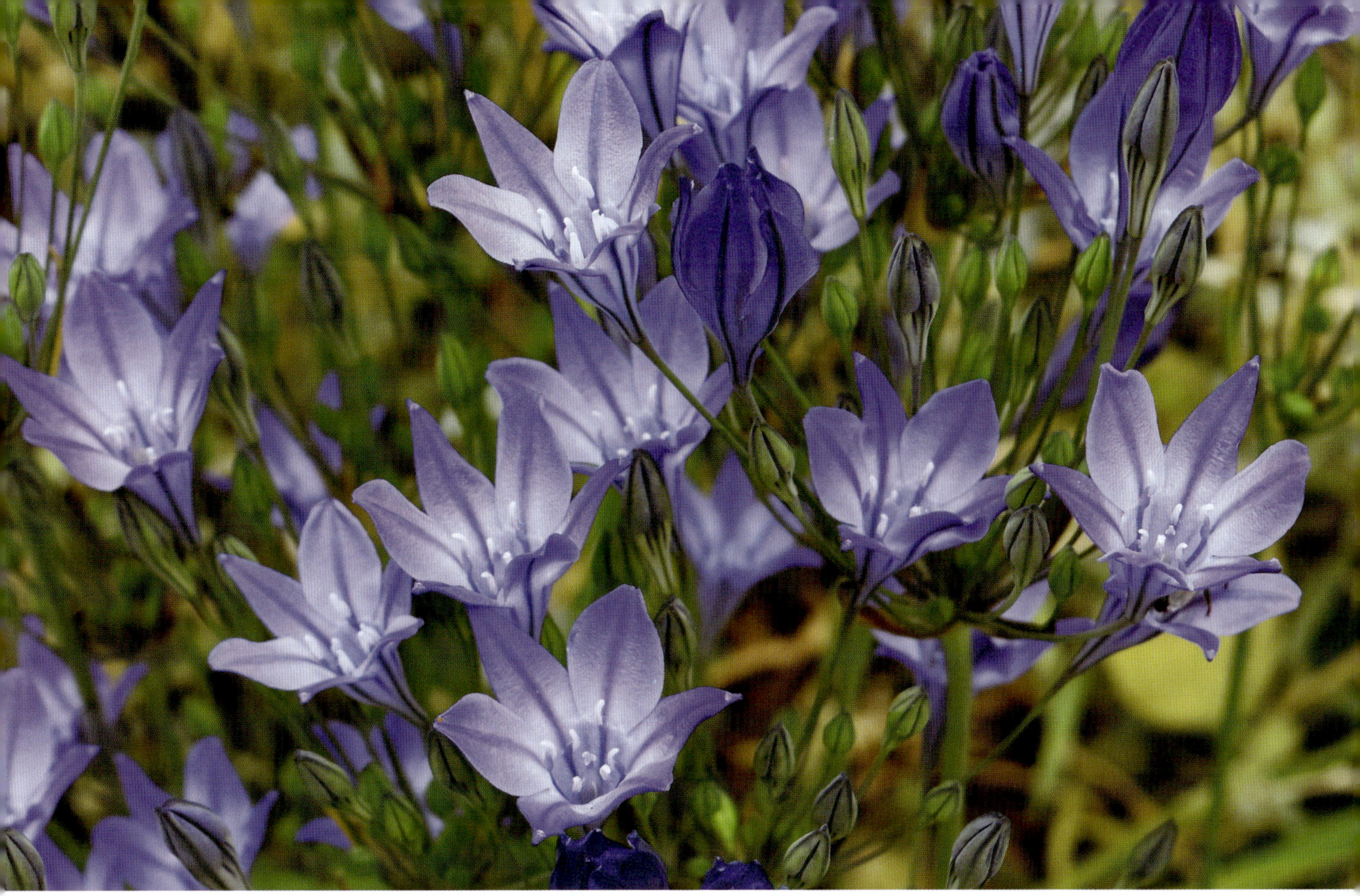

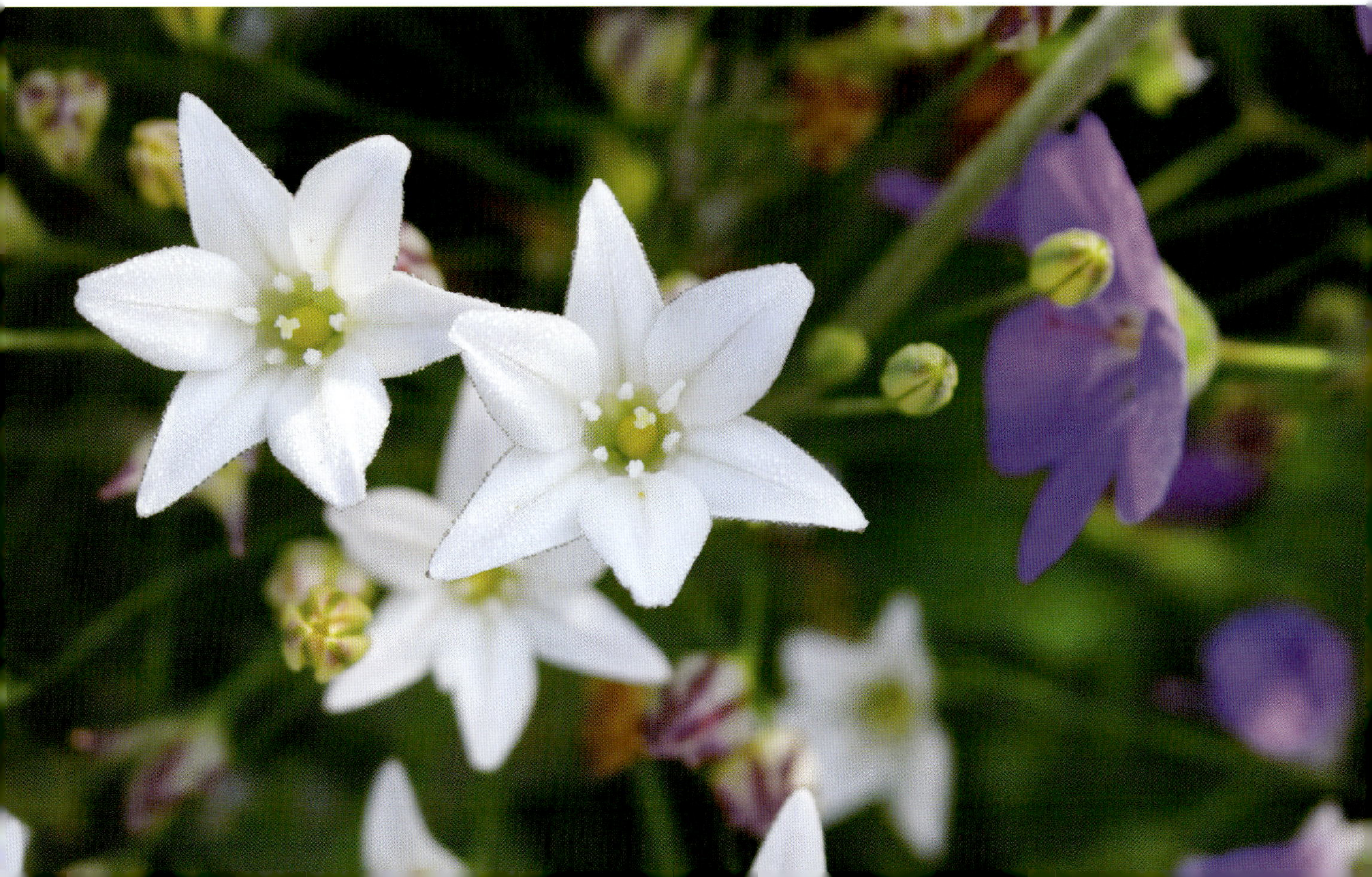

Tulipa

tulip

Full sun // Rich, well-drained soil // Zones 3–9 // Height 4–30 in. // Season of interest—spring

About These Bulbs

Tulips are the darlings of the spring flower garden due to their beauty and variety. The most familiar tulips resemble cups or goblets in shape. The petals curve in at the top when in bud, and then gradually open up in sunshine to become star-shaped. Other less well-known tulip shapes are the multi-petalled doubles that resemble a full peony when in bloom; the lily-flowered tulips that have pointed tops that flare out; and the irregularly shaped parrot tulips. Tulip flowers are available in a paintbox of colors with every hue apart from a true blue. They can have one solid color or diverse patterns, such as contrasting or complementary bases, edges, stripes, or feathering.

The big, bold tulips that are seen in many gardens today are the descendants of original wild tulips that still grow in rocky, inhospitable mountain areas. Breeders have taken the original tulip species and crossed them to produce an incredible diversity of shapes, colors, and heights. There is a tulip for every garden. With a little thought, you can choose one or two that are right for you. When selecting suitable tulips, think about whether you are looking for tulips that you will treat as annuals, or ones that you would like to come back for a few years.

Planting and Maintenance

The ideal conditions for tulips are places with six or more hours of sunshine a day, in well-drained soil. Plant them in the autumn in mixed flower beds, raised areas, or in containers. If you use tulips in your garden for maximum show, you will want to buy and plant new bulbs each fall. With this technique, you are getting top-sized bulbs that have been prepared by the grower to be vigorous and floriferous. After they flower, you can pull the bulbs out of the ground and compost them—or dispose of them if they look diseased. The same bed or container can be used for the next seasonal display. In warm winter climates, purchase cold-treated tulips that can be planted 8 to 10 weeks before you want them to bloom.

Plant small tulips as soon as you receive them, but you may want to wait to plant hybrid tulips until late fall. A fungal disease called tulip fire, which causes mottling and streaking of the petal

Tulips come in a wide range of colors.

colors and distorted browned foliage, can affect tulips that are planted into soil that is still warm. Planting into cold soil seems to help reduce this problem. If you have the space, change the areas where you plant your bedded-out tulips from year to year. Ideally, you will not grow tulips in that same soil for three or four years.

Tulips are prone to damage by many predators, so take precautions, like growing them in fenced areas or in pots to reduce damage. There are sprays that you can use that have odors that animals find offensive. These will need to be reapplied after heavy rains.

Design Ideas

When planning your tulip show, the first decision is whether it is an annual or perennial display. The showy hybrid tulips are the classic stars of the spring garden. They bloom in late spring in an incredible range of sizes, shapes, and hues which can dictate the rest of the supporting flowers that you choose to accompany them. Growing tulips is an ideal opportunity to indulge your color senses to the maximum. The good news is that because most gaudy tulips are grown as annuals, you can experiment this year and try again next year.

If you are deciding on a combination of different tulips in an area, think about whether you want all of your tulips to bloom in one big show, or for them to flower over a period of time. The most spectacular look comes from having them bloom all at once. This allows you to come up with color schemes, shape contrasts, or whatever else you can dream up. The tulip season will overlap with daffodils and other spring bulbs, which allows you further chances to make combinations. If you

prefer a progression of bloom, choose some tulips that bloom at different times: early, middle, and late season. Select tulips that look lovely together, in case they overlap in bloom time.

If you are using tulips in a mixed flower bed, then you will need to consider what plants to combine with them. Tulips are best interplanted with perennials that can cover the space when the tulips are finished. Try perennials with scented foliage such as herbs, catmint (*Nepeta*), or calamint (*Calamintha*) to disguise the bulbs. There are two benefits. The scent of herbs confuses browsing animals and the expanding perennial foliage covers the site where the tulips are planted. To dissuade herbivores from approaching the flower bed, plant strongly smelling alliums and crown imperial fritillaries.

As you plan where to plant your tulips, consider the heights of each type. The tallest tulips can grow to several feet, while the tiniest flowers hug the ground. Most tulips fit somewhere in between these two extremes. For a kind of tulip that does not look like a tulip, choose the coronet tulips, which have twisted petals that look a little like their namesake crown.

Tulips are categorized into divisions. Each group contains tulips that share a similar look or common heritage.

Division 1, single early tulips, are some of the first to bloom in early to mid-spring. Their short stems are just about a foot in height. They are well suited for growing packed together in annual bedding, situated at the mid-front of the flower bed, or in containers. Some of these tulips are fragrant. These are useful for windy coastal or hillside gardens, due to their short stems. This division includes some of the oldest cultivars.

CLOCKWISE FROM LEFT: *Tulipa* 'Keizerskroon' // *Tulipa* 'Apricot Beauty' // *Tulipa* 'Foxy Foxtrot' // *Tulipa* 'Dressing' // *Tulipa* 'Synaeda Blue' // *Tulipa* 'Pretty Princess'

Bright red-and-yellow 'Keizerskroon' dates back to the 1750s. 'Apricot Beauty' and 'Christmas Dream' are other popular cultivars. This division is useful for forcing into early bloom. Their short stems do not flop as much as taller tulips.

Division 2, double early tulips, are similar in stature and bloom time as the single early tulips. Their wide bowl-like flowers are filled with extra petals. In the garden, interplant double early tulips with other low-growing bulbs, or pack them in tightly for maximum impact. Some of the double earlies are fragrant, like the variably yellow 'Foxy Foxtrot' and brilliant red 'Abba'. 'Monsella' is a fun yellow with red flame-like streaks that I have grown. 'Dressing' is a lovely soft pink with some green on the outer petals. This group makes great container plants and forces well.

Division 3, triumph tulips, are one of the most popular of the tulip groups for their strong straight stems, midseason blooms, and brilliant colors. They are usually around one and a half feet tall and are excellent for forcing into early bloom in pots. Popular cultivars include 'Negrita', with rich purple flowers, and old favorites like the deep red 'Couleur Cardinal' and 'Prinses Irene', which is orange with a purple wash. 'Synaeda Blue' is not really blue, but plum-purple, with lovely white edges to the petals. For variegated foliage, try the bright and light pink flowered 'Pretty Princess'.

Division 4, Darwin hybrid tulips, are grown for their long stems and large, brightly colored flowers in mid to late spring. These tulips may perennialize in the garden if they are planted deeply and allowed to die back fully after flowering. Darwin hybrids are also used in tulip bedding

Tulipa 'Light and Dreamy' // *Tulipa* 'Beauty of Spring' // *Tulipa* 'Queen of Night' // *Tulipa* 'Flaming Club'

schemes, due to their wide range of stunning colors and strong stems. The flowers are egg-shaped when closed. When they open, you can see the contrasting color blotch inside at the base. There are lots of cultivars in this division that make good garden plants. The Impression Series is well worth growing. I have grown 'Apricot Impression' and want to grow 'Design Impression' now that I have seen it in person. I love the colors of the cultivar 'Light and Dreamy', which has pinky lilac flowers and strong dark stems. 'Beauty of Spring' lives up to its name with yellow-cream petals and red edges.

Division 5, single late tulips, are tall late spring wonders that are widely grown for their reliable blooms. Their flowers have that elegant, classic tulip shape and are used extensively in the garden and the vase. They have long stems, to two feet or more, that make them good choices for growing up through lower spring perennials. Single late tulips are also an excellent choice for growing in warmer climates because they respond well to pre-chilling. Plenty of colors are available in this group. Some of the top ones are 'Maureen', which changes from cream to refined white; 'Menton', which has shades of pink through apricot; and the silky, dark maroon 'Queen of Night'. For something different, try 'Flaming Club', which produces several flowers per stem, each with deep pink-red-and-white stripes. If you are a tulip lover, you need to include at least one of this group on your wish list.

Division 6, lily-flowered tulips, are easily recognizable with their sharply pointed petals that flare out at the top when closed and look star-shaped when they open in the sun. Lily-flowered tulips are

CLOCKWISE FROM TOP LEFT: *Tulipa* 'West Point' // *Tulipa* 'Healthcare' // *Tulipa* 'Color Fusion' // *Tulipa* 'Spring Green' // *Tulipa* 'Honeymoon' // *Tulipa* 'Virichic'

elegant in the garden and in the vase. Stem length is between one and a half to two feet. Bloom time is mid to late spring. Some are fragrant. Try the distinctive 'West Point' for a clear yellow, 'White Triumphator' or 'Healthcare' for pure white, or 'Mariette' for a color-changing pinky rose with lighter edges and base.

Division 7, fringed tulips, have soft petal margins that look like ice crystals when seen up close. Colors and plant height vary according to the parentage. They bloom in mid to late spring. The flowers are mostly singles, but there are some doubles like the rose pink-and-white 'Queensland'. For singles, look for 'Lambada', which has a purple base that blends to a butterscotch top and spines; 'Honeymoon' for a lovely fresh white; 'Louvre', which has a dark purple-blue base that shades to light purple and an even lighter fringe; or 'Color

Fusion' for its creamy yellow petals, raspberry pink fringes, and black centers.

Division 8, viridiflora tulips, have petals that are feathered or streaked with green in the centers. This distinctive look is a fun addition to any planting scheme, especially if you match the outer color to another plain color tulip in the same hue. They bloom late in the season and hold up well in the garden and as a cut flower. Their height varies by cultivar from a foot to nearly two feet. One of my favorites, 'Virichic', has lovely pink-and-green petals. 'Spring Green', which is a creamy white and green, is a subtle and elegant choice.

Division 9, Rembrandt tulips, are historic tulips that were loved during the Dutch bulb mania. Red-and-white tulips like 'Silver Standard', which dates from 1760, have feathered or flame-like

CLOCKWISE FROM TOP LEFT: *Tulipa* 'Silver Standard' // *Tulipa* 'Grand Perfection' // *Tulipa* 'Black Parrot' // *Tulipa* 'Flower Power' // *Tulipa* 'Queensday' // *Tulipa* 'Negrita Double'

colored patterns that people found extremely attractive. They did not know that these variable patterns were caused by a tulip virus. Virus-affected tulip bulbs are no longer sold to gardeners. Instead, there are some tulips that have the same look but do not have the virus. They are classified in other tulip divisions according to their shape and heritage. If you want the look of these old-fashioned beauties, try modern Rembrandt tulip lookalike 'Grand Perfection', or buy a Rembrandt mix.

Division 10, parrot tulips, are late-blooming tulips that are unlike any other group. The colors and patterns may vary even within blooms of the same cultivar. The simplest one is the single-colored, dark-flowered 'Black Parrot'. The most outrageous ones, noted for their vibrant feathered colors and twisted petals, include 'Estella Rijnveld' and 'Flaming Parrot'. Flower

Power' is a great orange with green and yellow markings. 'Apricot Parrot', one of my longtime favorites, has shaded petals that include some green in the centers.

Division 11, double late tulips, are not the traditional ovoid shape. They have extra layers of petals that give the flower the look of a small peony on strong stems. In fact, they are sometimes referred to as peony-flowering tulips. They have a longer flowering period than single tulips. Double late tulips are great in beds but look pretty in pots too. Look for the dark-flowered 'Negrita Double'. The bright and cheerful 'Carnaval de Nice' is striped red and white. 'La Belle Epoque' and 'Copper Image' both have subtle combinations of coral, copper, and apricot. Choose 'Queensday' for a fragrant, rich orange flower with green shadings, and a fine yellow line at the edges.

CLOCKWISE FROM TOP LEFT: *Tulipa* 'Whisper' // *Tulipa* 'Bonbini' // *Tulipa* 'Purissima' // *Tulipa* 'Sweetheart' // *Tulipa* 'Red Torch' // *Tulipa* 'Quebec'

Division 12, kaufmanniana tulips, are short tulips that reach just 6 to 12 inches. They are some of the earliest to bloom. When the buds emerge, their shape is similar to other tulips, but as they mature, the flowers open in the sunshine and resemble a waterlily in shape. The colors are often arranged in concentric rings that look rather like a bull's-eye with a dark center. An early spring one is 'Heart's Delight', which has green leaves with purply lines and dots. The flower is mid-pink with a yellow flush at the bottom and a lighter pink inside. 'Whisper' is an old-fashioned tulip that has a similar flower but with a red-and-black center. 'Bonbini' is yellow with some red outside. 'Ice Stick' has slender white flowers with a yellow center and a blush of pink on the outer petals.

Division 13, fosteriana or emperor tulips, are early bloomers with large flowers that may be marked with black. They are borne on long stems. They are nice garden plants with wide leaves that may be plain or have markings. 'Albert Heijn' has pink flowers; 'Candela' is soft yellow with green leaves; 'Purissima', one of the best of this group, has flowers that open creamy yellow and fade to white; 'Flaming Purissima' has light and dark pink flowers that change color as they get older with green foliage; 'Sweetheart' is a two-tone yellow and white; and 'Red Emperor' is a classic red with a mustard-colored base.

Division 14, greigii tulips, are known for their short stems and highly marked leaves that are borne low to the ground. They are useful for containers, in rock gardens, as bedding plants, and in windy sites where their short stems are unlikely to get broken. The flowers open wide in the sun to give a full look. 'Red Riding Hood'

Tulipa bakeri 'Lilac Wonder' // *Tulipa clusiana* 'Cynthia' // *Tulipa* 'Little Beauty'

is bright red with green foliage striped with purple. 'Toronto' has reddish flowers with black and yellow in the middle and mottled leaves. Multiflowered 'Quebec' is red and creamy yellow. 'Red Torch', a lovely historic tulip, has shiny scarlet red flowers.

Species tulips are the original wild tulips and their cultivars. They are less well-known than the large, bold tulips, but they have their own sweet charms. Many of them need really well-drained soil in a sunny spot, but some can take part shade. All species tulips will perennialize and come back for several years if they are planted in a suitable location. Most of these bulbs need a summer baking, so clear away organic matter such as deciduous leaves that fall on the flower bed.

Species tulips are small in comparison to the hybrid ones. Plant them in a rock garden or as one type per container for the best chance of success. Plant species tulips to come up through low-growing creeping phlox (*Phlox*), hardy geraniums (*Geranium*), or other groundcovers that hold the little bulbs in place and protect them when they are in bloom. For easy viewing, tuck these slender, diminutive beauties beside a path, on a raised bank, or in a rock garden. The flowers open in the sun and then close in cloudy conditions and at night.

There are many lovely species tulips that grow well in gardens and are easily slipped into small areas. The colors, heights, and requirements vary, so check the size and growing conditions that they need before purchasing

Tulipa humilis var. *pulchella* Albocaerulea Oculata Group // *Tulipa sylvestris* // *Tulipa schrenkii*

them. *Tulipa bakeri* 'Lilac Wonder' has diminutive lilac-pink flowers with a yellow-and-white center. *T. clusiana* has several good cultivars, such as the red-and-yellow 'Cynthia' and the red-and-white 'Lady Jane'. *T. cretica* 'Hilde' is multiflowered and has white-and-pink flowers with a yellow center. 'Danique' has a pink-beige cast and a yellow center. Light yellow 'Honky Tonk', a favorite of mine, is about eight inches tall. 'Little Beauty' is bright pink with a blue-and-black base. *T. humilis* var. *pulchella* Albocaerulea Oculata Group has a lovely coloring of white petals and a purple-blue base. Plant this tulip in some shade to preserve its coloring. *T. praestans* 'Shogun' is a light orange which looks good in containers. *T. schrenkii* is a short and variable rock garden species that only grows to six inches.

T. sylvestris is yellow with a greenish yellow outside and will grow in full sun or part shade. *T. tarda* has petals that are mainly yellow but have white tips. *T. orphanidea* Whittallii Group has delightful orange flowers that I paired with the russet trunk of a paperbark maple (*Acer griseum*).

Spring-Planted Bulbs That Bloom in Summer and Autumn

The second part of the gardening year is prime time for summer bulbs. These are planted in the garden or in pots, after spring temperatures have warmed up. In the summer, they produce blooms and lush leaves that enliven our gardens. Additionally, there are unconventional bulbs that need planting in summer, which will bloom soon afterward, like autumn crocuses and colchicums. Others, like lycoris, are best planted in late spring to flower in summer.

As days get shorter and the temperatures get colder, it is time to assess whether your summer bulbs will survive outside in the coming winter. Bulbs that are hardy can be left in the soil, as long as they do not get waterlogged. If they are borderline hardy, they can be protected from cold with a layer of mulch on the soil surface. Tender bulbs will need to be dug up and brought in for the winter. If you do not want them next year, they can be sent to the compost heap.

As you decide which summer bulbs to include in your garden, think about whether you want to plant them in the ground or in containers. Let their shapes, heights, and colors inform your purchasing decisions. You will find some amazing bulbs that will flourish in the heat of the summer and brighten up your garden.

An apricot-colored dahlia is the centerpiece of this garden.

Agapanthus

agapanthus, lily of the Nile

Full sun to part shade // Moist, well-drained soil // Zones 7–11 // Height 12–48 in. // Season of interest—summer

About This Bulb

Agapanthus is grown for its mid to late summer loose balls of blue, white, purple, or bicolor tubular flowers that are clustered atop a tall, flexible stem. The number of flowers in the cluster varies, but there are usually dozens. Their strap-shaped foliage is pleasantly green and may be evergreen or deciduous, depending on the species or cultivar. For outdoor growing, deciduous types are hardier than evergreen ones. Agapanthus make good garden plants in Mediterranean climates, but they are grown in containers in many areas of the world. If you garden in a cool climate, place the pot against your hottest wall, so heat and light reflect back onto the plant. Agapanthus can be used as cut flowers in arrangements.

Planting and Maintenance

Depending on your climate, you may be able to grow agapanthus in a pot or in the ground. Before buying a particular species or cultivar, check the hardiness zones. In the ground, plant the bulbs a couple of inches below the soil surface, in full sun, and with well-drained soil. They require a summer baking to flower well. Agapanthus may grow better in containers if your garden soil is wet, even though they would theoretically survive in the ground.

In a container, use a well-drained potting soil with extra grit mixed into the soil. Agapanthus is widely used as a container plant, where it seems to react well to having its roots restricted. If you are planting it into a new pot, leave only about a finger's width of space between the roots and the inside wall of the pot. Let the agapanthus gradually fill the pot with roots before dividing again. Agapanthus is usually grown by itself in a pot because of its specific watering needs. Use terracotta pots to improve drainage. Regular watering is necessary during the growing period, in the spring through fall. Use a weak solution of fertilizer every couple of weeks. In the winter, the plant needs a dormant period when watering is much reduced and cool temperatures are required for best growth the following year. You could plant several agapanthus bulbs together in a large container. Plant the shorter cultivars closer together than the tall ones.

After flowering has finished, cut the long stalks back to the base, but leave the wide, strappy

Agapanthus Headbourne hybrids // *Agapanthus* 'Ever Twilight' // *Agapanthus* 'Black Jack'

leaves to provide energy for next year's bloom. Once the plant is pushing itself out of the ground or pot, it is time to divide the thick roots. This is a tough job and requires significant strength and, maybe, the help of a friend.

Design Ideas

Agapanthus plants flower in summer, when the blue color is most welcome and fresh among many other hot colors. Where they are hardy, they can be used in masses or interspersed as part of a mixed flower bed. In colder regions, agapanthus are wonderful plants for large containers on a summer patio or deck. Choose a short-statured agapanthus for small containers. Their foliage remains attractive even when the plant is not in flower.

Species and Cultivars

When choosing agapanthus, look at their hardiness zones. There are some that are hardy to zone 6 like the Headbourne hybrids. There are many new cultivars that vary in height and flower color. For short plants, choose dark blue 'Lilliput' or mid-blue 'Peter Pan'. For unusual colors, try 'Twister' or 'Ever Twilight', which are white and indigo blue; 'Storm Cloud', which is dark violet-purple; or 'Black Jack', which is deep purple and perfect for container growing.

Alocasia

alocasia, elephant's ear

Part shade // Moist, well-drained soil // Zones 9–12 // Height 24–120 in. // Season of interest—summer into fall

About This Bulb

Alocasias have marvelous, arrowhead-shaped leaves that you can use to decorate your summer garden. The obvious veins in the leaves are part of the showy display. Leaf colors can include green, bronze, purple, and nearly black. They are native to tropical parts of Asia, so they perform best in areas of high temperatures, humidity, and plenty of water. They need year-round heat and are best always kept above 60 degrees F. Some gardeners like to grow these bulbs as houseplants, keeping them indoors all year. Alocasias are generally resistant to herbivore browsing.

Planting and Maintenance

Alocasias grow from a spring-planted bulb. Plant the enormous bulbs four to eight inches deep in rich soil with the pointed tip upward. Choose a position out of strong winds, so that the leaves do not get damaged. They need part shade for good growth; if planted in hot sun, the leaves will burn and turn brown. Where you have a short growing season, start the bulbs into growth inside. They can be brought outside after the temperature has warmed. To get them to start growing, you may need to place the pots over a radiator or on a heat mat.

In hot climates, they can be grown outside all year. If you garden in areas that get cold winters, alocasias are best grown in containers that can be brought inside for the winter to a relatively warm spot. In the spring, wait for the soil and air to warm up before bringing them outside again. To grow large, these plants need lots of water. They are often grown near a pond or other water feature, where the overflow or splashing helps keep them watered. If you do not have this type of situation, then make sure that they are within reach of a hose.

Alocasias need plenty of nutrients in the soil to grow well. Add extra compost to the soil surface after planting. Once growth has started, regularly add a general multipurpose fertilizer every few weeks. The plant will require all major nutrients, including nitrogen, to grow its enormous leaves.

At the end of the growing season, bring your alocasias inside as soon as temperatures begin to cool down at night. Make sure that you bring them inside before they get frosted. Store them as bulbs in a semi-moist state, or, if you have the room, overwinter them as large houseplants. They can be kept slightly drier when they are inside because their growth rate slows down. Their leaves

Alocasia ×amazonica 'Polly' // *Alocasia* 'Sarian'

may fall off if you reduce the amount of water, but they will regrow when you start watering again next spring.

Design Ideas

Alocasias are a great addition to a shady sitting area. Their tropical-looking foliage acts as a temporary screen, or focal point. Combine them with other bold foliage plants like colocasias, caladiums, and some cannas if you have a nearby sunny corner. Add a few pots of tuberous begonias for extra color.

Species and Cultivars

Alocasia hybrids have a variety of different marking patterns, colorations, and leaf sizes, so choose your species and cultivars according to what look you want. They are generally grown for their leaf colors. There are some lovely choices with distinct light-colored veins on a dark green background, such as *Alocasia ×amazonica* 'Polly' and 'Mandalay'. The cultivar 'Sarian' has wonderful wavy leaf margins and deep green leaves. Others have dark burgundy and dark green foliage, such as the hybrids 'Metalhead' and 'Sumo'. If you want enormous upright leaves, then choose the giant taro, *A. macrorrhizos*, which has ruffled green leaves and prominent ridged veins. In ideal growing conditions, giant taro can grow 12 to 15 feet tall, where it is hardy. When grown in a container, like a large half wine barrel or equivalent, it does not reach the same proportions, even with plentiful watering. 'Borneo Giant' is one of the largest selections, as its name suggests.

Begonia

begonia

Part shade to full shade // Moist not wet soil // Zones vary //
Height 12–18 in. // Season of interest—summer into fall

About This Bulb

Begonias are one of the best shade-loving summer bulbs, with their decorative leaves and single or double ruffled flowers. They are available in a wide range of pastels, bright colors, and bicolors. Begonias can be grown in the ground for bedding displays, path edgings, and around patios. They are most often used in containers, where their delicate stems can be protected from breaking, and their features can be seen clearly. Some begonias are upright, while others are best around the edges of the pots so they can drape over the rims. Not all begonias grow from bulbs. The ones mentioned here are just a few of the possible types that are available, some of which are usually grown as houseplants.

Planting and Maintenance

Begonia bulbs should be planted with the concave, sunken side facing up, about an inch below the soil surface. This is where the shoots and flowers will emerge. The roots come out of the convex side of the corm. These begonias can be slow to begin growth, so it is best to start them off inside in early spring. Transfer them outside once all frosts have finished. Give them moist soil but do not let them sit in water. During the growing season, take off the old flowers to keep the plants looking fresh.

In cold climates, most begonias are not hardy in the ground. They can be dug up before the first frost and saved inside over the winter. If they are grown in pots, bring in the whole container and let the compost gradually dry down. The leaves will dry up and fall off. Keep the potting compost slightly moist in the resting phase. Begin watering as days warm and lengthen in the spring.

Design Ideas

Begonias are used in the ground alongside fellow shade-loving bulbs, like caladiums, and perennials, like hostas and ferns. In containers, they can be grown alone or added to mixed pots for shade. They look particularly good in window boxes and hanging baskets. Some begonias have a cascading habit. They look good planted at the edge of pots, so they can dangle down. The most important thing when choosing your begonias is to look for ones that are suited to your climate. The large-flowered *Begonia ×tuberhybrida* is a wonderful choice where summers are mild,

Begonia 'Fragrant Falls Peach' // *Begonia* 'Lemon Berry' // *Begonia* 'Crispa Marginata'

whereas *B. boliviensis* and its cultivars hold up well in warmer temperatures. Also, choose begonias by the color of the leaves and flowers, so that they will coordinate with surrounding plantings.

Species and Cultivars

Begonia × tuberhybrida is available in a wide range of colors, including red, yellow, white, and orange. It has bold flowers that are great for large pots and hanging baskets. Plant a mixed selection to brighten up a shady corner, or choose one type if you have a set color scheme. Some cultivars are scented, such as the Fragrant Falls Series.

Other tuberous begonias make great additions to summer containers for partly shady areas. They will need regular watering and fertilizing to promote a long season of bloom. They can be dug up and saved for next spring, or the whole pot can be brought inside for winter. Most are only hardy in zones 10 and 11.

Begonia 'Crispa Marginata' Group are a tuberous begonias. This group contains some lovely begonias that have contrasting colored edges to their flowers. Flower color may be peach, yellow, or white with a red margin. They grow to about eight inches tall and are fantastic for containers, where their large flowers are shown to best effect. Plant them in part shade. They are sometimes sold as single colors, but are often available as mixes.

Begonia Nonstop Series are tuberous begonias that give a long season of bloom, in a range of colors from white and yellow through oranges and reds. Their main benefit is that deadheading is not needed, and, as their name suggests,

Begonia 'Nonstop Fire' // *Begonia* 'Glowing Embers' // *Begonia boliviensis* 'Mistral Yellow' // *Begonia grandis*

they keep on blooming all summer. Plant them in containers for a shady show.

Begonia 'Glowing Embers' is a tuberous begonia with a trailing habit that lends itself to hanging baskets and window boxes. The plants are single and vibrant orange in color. The green leaves have a contrasting color on their underside.

Begonia boliviensis is one of the best begonia species to grow where you have hot and humid summers. It grows one to two feet with a pendulous appearance. Look for white 'Santa Barbara', with green leaves; light yellow 'Mistral Yellow'; warm pink 'Mistral Pink'; red 'Santa Cruz', with dark green leaves; and the slightly salmon-red 'San Francisco'.

Begonia grandis is a shade-loving hardy begonia that grows well in zones 6 to 10. The arching stems are one and a half feet long, and bear pink or white flowers in late summer or early autumn. The wing-shaped greenish leaves have red underneath. It emerges late in the spring and looks great as a path edging or groundcover. It performs well in moisture-retentive soil. I pair *B. grandis* in the same spot as Virginia bluebells (*Mertensia virginica*). The bluebells grow up first and, as they are dying down, the hardy begonia takes over the show for summer into fall.

Caladium

caladium

Full shade to part sun // Rich, moist, well-drained soil // Zones 10–11 // Height 12–24 in. // Season of interest—summer into fall

About This Bulb

Caladiums are grown for their long-lasting, heart- or arrow-shaped leaves with colorful patterns that can brighten up a shady part of your summer garden. Some new caladiums have been bred to grow in partly sunny places, as long as the soil retains plenty of moisture.

They work well in the ground or in a container. Caladiums come from warm to hot areas, so they grow when the soil and air temperatures are warm. In hot climates, caladiums are a year-round addition to the flower bed. Where winters are cold, their leaves are a cheerful garden presence all summer. Treat caladiums as an annual in colder climates, or save them inside over winter, and bring them out again next year.

Look at the ultimate size and height of the plant. Caladiums range from small ones for little pots or the front of the border, to bold leaves that fill a large container. Next, think about the color scheme that you are planning. The colors of caladiums are reddish pink, maroon, chartreuse-green, dark green, white, and cream. The color patterns follow the leaf veins with a dominant central patch of color. The middle might be surrounded by a contrasting color and then edged in a different one. Other caladium cultivars have leaves with one central color and an edging band of a dissimilar one. In addition, there are likely to be spots and blotches of one of the colors all over the leaf. If this all sounds too much for you, then choose one of the more restrained color combinations.

Lastly, look for caladiums with elongated leaf shapes, frilly leaves, or those that are taller than regular caladiums. Some of these are good choices for hanging baskets. Plant breeders are producing a dizzying array of cheerful caladiums, so there is something for every garden.

Planting and Maintenance

Grow caladiums in organic-rich soil that has good drainage in a site with dappled shade. Caladiums are not tolerant of drought. They require regular deep watering throughout the growing season. Caladiums are usually grown as shade plants, but some have been bred for use in sun. Look for cultivars in the Florida Series. Because the leaves are quite thin, I would provide caladiums with morning sun and afternoon shade in hot climates. Part sun helps to develop the leaf color, but the colors fade and the sun can scald the leaves in too much light.

If you live in a climate with short summers, it is helpful to start your caladium bulbs inside, about six to eight weeks before your last frost. Place the tubers in trays of potting compost in a warm area. Cover them with about an inch of potting mix. Water them well and place them in a bright windowsill or under lights, preferably with bottom heat on a heat mat or radiator. They are sometimes slow to start into growth, so this gives you a few more weeks of good growth outside. This is especially helpful when planting up containers. You can then put out a plant that is already growing.

When planting caladium bulbs directly into the ground, wait until the day and night temperatures have warmed up. The bulbs will rot and fail if they are put outside when the soil is still cold or wet. Dig a wide hole about four inches deep and add in a couple of inches of organic material. Plant the bulb about two to three inches deep. In sandy soil, add more compost and plant at about four inches, so that they get enough moisture at their roots. Add organic matter and soil to cover the bulb. As the plant grows, backfill the hole with humus-rich soil.

Caladiums require regular watering to make new leaves all summer. If your soil is poor, add plenty of organic matter to the soil surface. You could use some liquid fertilizer in your watering can every few weeks as they are starting into growth. They are leafy plants, so look for a nitrogen-based fertilizer. Ease off on the watering and fertilizing late in the season; you want to encourage your plants to begin to shut down top growth and store energy in their bulbs.

Late in the summer, the leaves begin to die back as cool autumn temperatures arrive. This is the time to bring the caladiums inside. If frost has nipped the foliage, cut them off, leaving part of the stem. Dig up the bulbs and bring them into a dry frost-free area. Shake or brush off the soil and let any remaining leaves shrivel up, and then gently pull them off. Store them wrapped in several layers of newspaper, in a crate or box. In cold summer climates, caladium can be grown as a houseplant. If you are growing caladiums in a pot, the whole thing can be brought inside. They will not need regular watering during the winter, so they can be stored on their sides. They do best with a period of rest, during which time you give them little water. Bring on new leaves in spring by potting them up and gradually starting to water them again.

Design Ideas

Whatever leaf colors you choose, they will bring their unique personality to your flower bed. They are a fantastic addition to a shady space, but they cannot compete with surrounding roots, so plant them away from shrubs. If in doubt, do not plant them in the ground; instead, use a deep pot with moisture-retentive potting soil, and place the pot into the shady area as an accent. This also makes it easy to bring the caladiums inside for the winter.

They make great container plants as long as you can water them regularly. Place them around the edges of the pot, where the leaves can dangle down over the side to soften the rim. Plant only caladiums, or pair them with some shade-loving begonias for a change of shape and texture. Keep plants in a similar color palette for a cohesive look. For example, use red begonias with a red-hearted

Caladiums are one of the best bulbs for a shady summer container.

caladium, or choose white caladiums planted with light-colored summer annuals.

When they are planted in the ground, caladiums are perfect paired with other shade-loving tropical plants or hardy perennials. Make a planting pocket between existing hardy clumps of hostas or ferns and add caladiums there every spring. Those with predominantly white leaves add a soothing, peaceful look to a bed or container. For a white-and-green summer shade garden, combine them with white-variegated hostas, lily turf (*Liriope*), hardy gingers (*Asarum*), or ferns. Any of the cultivars look good lining a shady path. A sheltered or protected corner is a fabulous positioning for caladiums because they do not do well in a windy garden. The leaves whip around on their flexible stems. In a gentle breeze, their movement is pleasant, but the leaves can tear as soon as the wind picks up.

Species and Cultivars

Caladium ×hortulanum has been bred to produce a wide range of colors, leaf shapes, patterns, and speckles. You may just see it listed as *Caladium* followed by the name of the cultivar. Some of the most prized caladiums have white in their leaves so that they show up well in the shade. 'Candidum' is one of the classic caladiums that has been used in gardens for years. 'White Christmas' and 'Moonlight' are other white cultivars. 'Summer Breeze' is an elegant shade lover that has pink veins and green margins. For red-and-green leaves, choose 'Restless Heart', 'Blaze', or 'Red Flash', which is sun tolerant. For pink, try 'Apple Blossom', 'Pink Splash', and 'Fanny Munson', which is known for lovely pink leaves, dark pink veins, and an interesting green leaf margin. 'Lemon Blush' has a red-pink centered leaf with wide chartreuse-green margins. There are some intensely patterned leaves, like those of 'Wildfire', which has a dark pink center, white veins, and green margins, speckled and spotted with pink. 'Miss Muffet' is a small caladium with chartreuse, white, and pink leaves. A different color choice is the sun-loving, bronze-leaved 'Burning Heart'.

CLOCKWISE FROM LEFT: *Caladium* 'Summer Breeze' // *Caladium* 'Red Flash' // *Caladium* 'Burning Heart'

Canna

canna

Full sun // Moist to wet soil // Zones 7–11 // Height 24–104 in. // Season of interest—summer into fall

About This Bulb

Cannas are easy-to-grow plants that give an exciting presence to your summer beds and containers. They are grown wherever you need a bold tropical look. Their leaves are large, dramatic, and clustered into a vase shape. The tall flower spike emerges from the middle of the leaves and is topped with brightly colored flowers.

The leaves provide interest even before the summer flowers emerge. There is a central midrib and sometimes prominent veins that demarcate areas of different colors on the leaf. Leaf color may be green, blue-green, bronze, stripy orange, purple, yellow, white, or cream. The colors of the leaves and flowers often contrast. Flowers are predominantly bright colors with orange, salmon or peach, red, hot pink, yellow, and a few cooling creams. Some have speckles, spots, or several contrasting colors within the flower for added decoration. Pollinators, especially hummingbirds, love canna flowers. Cannas are rarely eaten by deer.

Planting and Maintenance

Cannas do best in areas with hot summers. They need full sun and plenty of water to fuel their rapid growth. If you are in a cool summer climate, grow cannas in a sheltered, warm, and sunny position. In very hot climates, some afternoon shade will be best for growth. Plant the bulbs, that are a rhizome, about four to six inches deep.

Cannas are generally large plants that require a fertile, organic-rich soil. During the growing season, especially early on when they are growing large leaves, they need copious amounts of water. To ensure that the soil does not dry out, plant them within reach of a hose, especially if you are growing cannas in pots. For an in-ground planting, consider a low-lying area of the garden or right by the downspout or rain barrel. Some cannas are best when grown in standing water, such as at the edge of a pond.

Cannas grow from a bulb that is planted outside when the air and soil temperatures are consistently warm. They may rot if they are planted in cold soil. Cannas can be started inside in spring in flats or containers of potting compost to give the plants a head start. If planting directly outside, dig holes four to six inches deep, and a couple of feet apart. Short cannas can be planted closer together than the massive ones that need their own large space. Put some rich compost in the bottom of the hole. Lay the bulbs horizontally and cover them with a few inches of organic-rich soil. Water thoroughly and continue to keep the soil moist.

If your cannas do not flower, they might need additional watering and feeding. As the cannas grow, top-dress the soil with organic material to help keep the soil moist. For cannas in containers, add a liquid feed to your watering can. Use a diluted organic fertilizer that has plenty of potash—a half-strength tomato fertilizer, for example—or cooled wood ashes. I like an organic feed like seaweed fertilizer or fish emulsion. You can leave a saucer under your canna plants to keep water around the roots.

During the growing season, spruce up the plant by removing any leaves that get damaged by wind or insects. Cut off individual blooms as they fade. Then, when there are no more buds left to open, remove the whole flower stalk to the base. If you are growing cannas primarily for their foliage, you can cut off the flower stalk early, especially if you do not like the color combination of foliage and flower. Cannas can get eaten by some summer insects like Japanese beetles that gnaw on the leaf edges, and other insects that eat holes through a rolled-up leaf (so when it opens it looks like you have used a hole puncher). Remove affected parts of the plant and hand squish the insects as necessary.

When autumnal frosts start to injure canna foliage, it is time to dig them up for winter storage. The large-leaved tall cannas put on a lot of top growth, so be prepared for a correspondingly big increase in bulb size. Chop around the base of the plant in a circle and lever the whole bulb out of the ground. Cut the foliage back, leaving part of the stem on the plant to allow you to lift sections out of the ground. Let the roots dry out in a place that does not freeze. Store the bulbs in a cool dry spot for the winter. Cover them with thick layers of newspaper, or another type of protective layer. Check regularly to see that they are moist but not wet.

If you divide your cannas and make new plants, each division must have one or more good buds. Sometimes they will fall apart in pieces, or you can see where there is a natural division and cut with a sharp, disinfected knife. Division can be done in the autumn or in the spring.

Design Ideas

Cannas add a tropical look to the garden with a bold texture that contrasts well with small foliaged plants. Their large leaves and often vibrantly colored flowers are used where you need something eye-catching. They can be a focal-point plant, a repeated element throughout a bed, or a summer screen. They look great around a water feature, where they relish the extra water and humidity. Pair cannas with summer annuals like bright zinnias and marigolds, and other tropical plants such as hedychiums. Choose your cannas by plant height, flower, and leaf color. Plant heights vary from one or two feet to a towering six to nine feet. Shorter cannas bloom before tall ones because they need to grow less leaf area than the large ones.

Where you have to dig them in the fall, include them in a bed where all of the plants need digging up for the winter. Another way is to plant cannas in containers that you can bring into a frost-free area as fall temperatures cool down. Small-sized cannas work best in containers. You can plant large cannas in containers, but you will need super-sized ones that have extra bottom weight so that they do not blow over. Half wine barrels work well with added stones in the bottom. Choose a site in a sheltered corner, out of the wind. Position the containers correctly before you plant them up because they are a bother to move after planting on account of their weight.

CLOCKWISE FROM LEFT: *Canna ×generalis* Cannova Bronze Scarlet // *Canna* 'Orange Punch' // Cannas are often grown for their interesting foliage. // *Canna ×ehemanii*

Species and Cultivars

Cannas come in a wide range of heights, leaf and flower colors, and forms. I have grown the red-and-yellow-flowered 'Cleopatra', the dark-leaved 'Australia', and the gigantic 'Musifolia', with its banana-like leaves. Many others are well worth trying, including the green-leaved 'Orange Punch', the stripey-leaved 'Pretoria', and 'Black Knight', with its scarlet red flowers and bronzy maroon foliage. For a slender flower on a tall plant with green leaves, choose the elegant *Canna ×ehemanii*. The Cannova Series includes cannas with different flower and foliage colors, from bright orange with green leaves to pale yellow, rich pink, and red flowers with dark leaves. They are grown for their strong colors and ability to flower even in areas with cooler summers. Another tall choice with bold leaves is the Tropicanna Series. The first year that you grow them, they may not be very tall. They are often taller in subsequent years when the bulb has stored more energy. For containers, choose cannas from the compact CannaSol Series, such as 'Happy Emily', which has green leaves and variably colored yellow flowers.

Colchicum

colchicum

Full sun to part sun // Moist, well-drained soil // Zones 4–9 // Height 6–12 in. // Season of interest—fall

About This Bulb

Colchicums are grown for their chalice-shaped autumn flowers that are similar in shape to a crocus. This gives rise to one of their common names, autumn crocus, which is confusing because they are not crocuses. Colchicums bloom in colors of lavender-purple, purply pink, and white, and can be singles or doubles. If you look closely at the flowers, they may have a tessellated pattern similar to some spring-blooming fritillaries. The showy flowers bloom at an unexpected time, so they really get noticed in the garden. They are not bothered by herbivores.

One of their distinguishing features is that colchicums bloom in late summer or early autumn without any leaves. After the flowers die down in late autumn, there is no sign of the plant until the broad, elongated, glossy leaves emerge in a clump in the spring. By that time, most gardeners have forgotten what they planted there, and the leaves that pop up are a mystery, especially if you have not grown colchicums before. The leaves die down in late spring and the spot is bare in summer.

Planting and Maintenance

Plant colchicum bulbs as soon as they are available, which is usually in late summer. Their peculiar lifecycle means that they are not widely available, so you may have to search them out. Plant them about four or five inches deep in rich, humusy soil. They need to have some moisture in late summer to bloom, but they will not do well when regularly irrigated. They are best in a position with plenty of sunshine, protected by the part shade of some deciduous trees or shrubs. When their leaves come up the following spring, make sure that they get adequate sun and moisture. Planting in or among surrounding low-growing plants is a great way to support the sometimes floppy flowers. It also fills in the space left behind after the leaves die down.

Design Ideas

Colchicums need careful placement in the garden. Often the best positions are those areas of the garden that are at the edges, near shrubs, or in wild sections. The alternating pattern of flowers at one time of year, and leaves at another, can be difficult to manage in a manicured mixed flower

A grouping of colchicum // *Colchicum autumnale* 'Album' // *Colchicum* 'Lilac Wonder'

bed. Look for an area of grass or low groundcover that could be enhanced by some lovely late-season blooms. Plant them in singles or small groups to pop up through the surrounding plants.

Species and Cultivars

There are many species and cultivars of colchicum, and they are all very easy to grow. The flower colors are either white or lavender-purple. Some flowers have marvelous, tessellated patterns. *Colchicum cilicicum* has purplish pink petals that are four to six inches tall. Each bulb makes multiple flowers. C. 'Waterlily' is a popular cultivar which has full-looking flowers with many petals. C. 'Lilac Wonder' is a good garden plant with large flowers in soft purply pink that is only hardy to zone 5. We love the look of this growing up through gravel mulch. I really enjoy white colchicums like C. *autumnale* 'Album' and C. *speciosum* 'Album', which look great when planted between green groundcovers because they show up nicely. The smaller flowers are very elegant with their bright yellow stamens.

OPPOSITE: *Colchicum cilicicum*

Colocasia

colocasia, elephant's ear

Sun to part shade // Moist to wet soil // Zones 9–11 // Height 36–72 in. // Season of interest—summer into fall

About This Bulb

Colocasia plants produce dramatic leaves on long stalks that are a delight in summer shade gardens. The leaf shape is like a heart or an arrowhead. They grow so large that they are sometimes called elephant's ears. To prevent confusion with other plants, I have defaulted to calling them colocasias instead.

By the end of the summer, the leaves of colocasias may each be three feet long. The prominent veins come to a point at the dip in the heart-shaped leaf, where the stem emerges. The smooth, satiny leaf texture is very soft-looking. They are similar to, and often confused with, alocasias. Colocasias leaves usually point downward, and, in some species, their veins are less prominent than alocasias. They have an insignificant pale yellow flower that is not prized as a decorative feature.

Planting and Maintenance

Colocasias require warm soil to do well. Choose firm, large bulbs, which will give you the best chance of success. They can be directly planted outside after the soil and air warm up. Dig a big hole and add extra organic matter to the bottom. The bulbs will need to be covered with four to six inches of rich soil. In gardens with a short summer, it is better to start the bulbs inside in containers in late winter or early spring. Grow them in a warm area of your house. Plant out into the garden when you would put out your tomatoes or warm-season annuals.

Situate your plants in an area of the garden that gets morning sun and bright dappled shade later in the day. Choose an area that is naturally wet, or one where you will be able to water regularly. In a relatively short time, these plants put on a lot of growth, and most of that is composed of water. If you have a pond, they can be placed in a container on a ledge, with their roots at least partially submerged in the water. In cold areas, the plants should be taken out of the pond in fall and stored inside for the winter. Colocasias also need to be in an area that is protected from wind. Wind gusts can rip the large leaves and knock over container-grown plants. A sheltered, partly shaded patio is ideal.

In areas where *Colocasia* is hardy in the ground, treat your tubers to a nice thick layer of organic mulch to protect them during the winter. This will break down during the next growing season and help enrich the soil and add some nutrients, but, more importantly, keep moisture in the soil. In cold winter areas, dig up

Colocasia esculenta 'Blue Hawaii' // *Colocasia esculenta* 'Diamond Head' // *Colocasia esculenta* 'Illustris'

colocasia plants as the growing season draws to a close. Cut off the tops, and let the roots dry out for a few days out of the cold. Wrap them up in newspaper or use shredded paper to keep them moist, but not wet. Place them in a frost-free area for the winter.

Design Ideas

Colocasias are dramatic plants that come into their own in the heat of summer and last until the first frost. The leaves are large and bold. They provide a great contrast to fine-textured hardy perennials and annuals. There is no obvious flower color to clash with, and the leaves provide a great backdrop for surrounding plants. The leaves are often green; however, some cultivars bring the same large leaves to the garden, but with other desirable characteristics like a different leaf color.

These plants, which are native to tropical regions of Asia, need warm weather to do well in the garden. Where they are not hardy, they are often grown in containers so that they can be brought inside for the winter. They are used in conjunction with shade-loving plants that also like rich, moist soil, such as begonias, hostas, and ferns.

Species and Cultivars

Colocasia gigantea is the largest of the group. Its massive leaves will dwarf a small garden, but they are great where you have space to fill. With copious amounts of water and plenty of rich soil, the leaves can reach five feet long, especially where the plant can stay in the ground all year. 'Thai Giant' is the cultivar that is generally available.

Colocasia esculenta, taro, produces an edible tuber, which, when cooked, is used as a food called taro or dasheen in some cultures. The ones that are usually sold for garden use are large-leaved selections of this plant that are not used for food. Do not try to eat your plant if you are not an expert. The margins of the leaves have some waviness, and the veins are noticeable. A good group called the Royal Hawaiian Series has a variety of colors. 'Blue Hawaii' has mid-green leaves with dark-colored veins and stems, 'Diamond Head' has dark glossy leaves, and 'Illustris' has a green base with dark in between the veins. For large heart-shaped leaves that are dark purplish black, look for 'Black Coral', 'Black Ruffles', or 'Black Magic'. For golden green leaves choose 'Maui Gold'.

Crinum
crinum, swamp lilies

Full sun // Moist, rich soil // Zones 7–11 // Height 24–48 in. // Season of interest—summer

About This Bulb

Crinums come from tropical and subtropical regions. In cold climates, they need to be grown in pots or lifted from the ground in autumn. They bloom from summer into fall, depending on the species, with pink or white scented flowers. Crinum grow from big bulbs and can become large plants, so give them room to expand when you position them. The wide, strap-like leaves are evergreen in mild climates. The flowers are held at the top of upright stems in groups. Each bloom, which resembles a pink to white true lily, is fragrant when it opens late in the day. The plants and bulbs are resistant to being eaten.

Planting and Maintenance

In the wild, crinums, which are also known as swamp lilies, are found in marshy areas that are wet at certain times of year. In the garden, crinums benefit from a moist and rich soil, with added compost and leaf mold on the soil surface. However, swamp lilies are often found in old gardens where they have had no extra care. If you are on the edge of their hardiness zones, plant them deeply and mulch heavily. Alternatively, plant them in pots that are brought in for the winter. Crinums do not do well when disturbed—if you have to move them, they may not bloom the following year.

Design Ideas

Grow crinums in good soil, especially where there is some moisture. If they are not hardy, try them in a large container. Crinums look lovely on the sunny side of shrubs, especially those with dark green leaves that show off their light-colored blooms. Wherever you plant them, leave them space to grow.

Species and Cultivars

Crinum ×powellii is a hybrid crinum that is a good garden plant. It has pink or white fragrant flowers in summer that are borne on tall stems. The long, narrow leaves remain over winter, if your garden stays above freezing. While it performs best in warm to hot climates, it can grow in zones 6 to 11. Grow it in containers where it is not hardy.

Crinum 'Ellen Bosanquet' is a popular hybrid that is hardy in zones 8 to 10. The deep

Crinum 'Ellen Bosanquet' // ×*Amarcrinum*

pink, fragrant flowers bloom in the summer. The plant is about two to three feet tall. Where this plant is hardy, it will form large clumps over time. The flowers can be cut for arrangements.

×**Amarcrinum**, a hybrid between *Crinum* and *Amaryllis*, looks and acts in the garden like a crinum. The fragrant flowers resemble pink lilies, and multiple flowers grow on each stalk. Hardy in zones 8 to 10, this plant is wonderful for gardens that have hot summers. Grow this pest-resistant bulb in rich, moist soil and full sun. It does well when regularly watered and mulched.

Crocosmia
crocosmia, montbretia

Full sun to part shade // Rich, well-drained soil // Zones 6–9 //
Height 12–48 in. // Season of interest—summer into fall

About This Bulb

Crocosmias are easy-to-grow plants with tubular to star-shaped summer blooms that grow on gently arching stems. Flowers have six petals, in colors ranging from red-orange (the most commonly available) to pure orange and yellow. The long, sword-shaped foliage is usually bright green; some cultivars have bronzy or grayish green leaves. Crocosmias make nice clumps after they have been in the ground for a few years. They hold their good looks from late spring until fall, when the leaves gradually turn yellow. Crocosmias make good cut flowers both when fresh and as seedpods.

The brightly colored flowers open from the base of the spike to the tip, with a succession of blooms. The overall look of the flower from above is rather like a set of fascinating fishbones or a feather, with an alternate pattern along the stem. They are not flat like a fishbone skeleton but instead point upward along the stem. Crocosmias are a favorite flower of hummingbirds. They are listed as an invasive plant in some countries; check your local listings before planting.

Planting and Maintenance

Plant crocosmias in full sun or partial shade in soil that is average to rich and moist during the growing season. It cannot be wet when the corms are dormant in the winter. You may find corms for sale, or a potted *Crocosmia* that is already growing. Buying potted plants is helpful if you want to preview the flower color and plant height. The unpotted corms are usually cheaper than plants. Plant corms in a distinct clump for best impact.

In cold winter climates, wait until spring to plant the corms. If you think that they will be hardy in your garden, plant them in the autumn. Choose a sunny position with good soil. Plant them four to six inches deep and the same distance apart. Planting deeply will help the plants stay upright without the need for staking. Arrange them to make an attractive grouping. The plants will eventually grow together to make a lovely cluster of flowers and leaves.

Most crocosmias need full sun, but the bronze-leaved crocosmias need a little more shade, especially late in the day. Make sure you water the corms well initially and in dry spells. After flowering, let the foliage die down naturally to recharge the bulb for next year.

Crocosmia 'Lucifer' // Crocosmia masoniorum

The cultivars vary in hardiness, so check the growing zones before buying. If this is your first foray into crocosmia growing, pick the tried-and-true types, like the ever-popular red cultivar 'Lucifer'. If crocosmias are not hardy for you, grow them in pots alongside other non-hardy, summer-blooming bulbs like dahlias. If your chosen cultivar is borderline hardy, mulch heavily around the plant. This keeps the soil cool and moist in summer and insulates the bulbs from cold over winter.

Crocosmias can be divided when the clumps get congested. To do this, dig around the whole clump and lever it out of the ground. Pull apart the bulbs, which, in this case, are corms. This means that they wither away after growth and make new ones at the bottom for next year. Divide the grouping into a few smaller clusters, and compost the old shriveled corms that are at the bottom of the healthy ones. Replant the good corms in soil amended with compost, at the same level as they were originally growing.

Design Ideas

Crocosmias are a standout, even when not in flower. They are architectural-looking plants with their bold clumps of upright to

arching, green-to-bronze leaves. When in flower, crocosmias add zingy red, orange, or yellow to a summer flower bed. They look fabulous with summer bloomers like rudbeckias, echinacea, lilies, tall border sedums (*Hylotelephium*), catmint (*Nepeta*), and warm-season grasses. Place them in the correct section of the flower bed, depending on the flower height. Most work well in the mid-front of the border.

Species and Cultivars

Crocosmia 'Lucifer' has bright red blooms and grass green foliage. This is one of the most reliable and hardy crocosmias. It is widely available and grows well in most gardens. The fiery scarlet red blooms, held on branching, arching stems, stand out from a distance, especially when backed with the mid-tone greens of the strappy leaves. Tall flower stems rise above the pleated foliage. 'Lucifer' will take some shade and variable moisture levels.

Crocosmia masoniorum, giant montbretia, is a tall crocosmia that is three to four feet high. The flowers face upward on top of the flat sprays from summer through to autumn. This works well in mixed flower beds. Plant groupings of bulbs among agapanthus and daylilies (*Hemerocallis*).

Crocosmia ×crocosmiiflora is a great plant in mild, temperate climates that provides some late-season color in the flower garden. In gardens where it does well, it can be super aggressive or invasive, so check your local status before planting. Look for the deep yellow 'George Davison', which has green leaves. 'Emily Mckenzie' has relatively large bright orange flowers with a red center and leaves that have a bronze wash. 'Emberglow' has vivid red tubular flowers with a yellow center and bright green leaves. 'Solfatare' is an old cultivar with apricot-yellow flowers and bronzy leaves.

Crocosmia ×crocosmiiflora 'George Davison'

Dahlia

dahlia

Full sun // Rich, moist, well-drained soil // Zones 8–10 //
Height 12–96 in. // Season of interest—summer into fall

About This Bulb

Dahlias are the classic flowers of the late summer and early fall garden. They brighten up flower beds or large containers with their vibrant blooms, at a time when other herbaceous plants are slowing down. Dahlias continue blooming until frost, with the number of flowers dropping off as days get shorter and temperatures fall. While dahlias are primarily grown for their flowers, some cultivars are known for their near-black foliage rather than the standard green. Dark-leaved cultivars are popular in mixed flower beds.

Flower sizes for dahlias range from cute as a button, to pompoms, to the much-vaunted blooms the size of dinner plates. Colors and shapes show equal diversity. Flowers can be white, all possible pinks, maroon, plum, purple, burgundy, red, orange, yellow, peach, blends of colors, variegated, or various bicolors. The single flowers are daisy-shaped and symmetrical with colored outer petals and a yellow center. These are great flowers for pollinators. The double flowers are varied in shape but contain a mass of petals in the center.

Smaller-scale dahlia plants are great in large containers or near the front of a flower bed. Tall dahlias look good integrated into the middle or the back of flower beds. Tall plants take time to make lots of leaves to power the production of dinner plate–sized blooms. They are some of the last dahlias to produce flowers. Medium or tall dahlias are often grown in a separate area and used to produce cut flowers. I have grown dahlias for decades and cannot imagine my summer garden without them. It is so rewarding to be able to plant them in the ground in spring and harvest blooms by summer.

Planting and Maintenance

Dahlias are easy to grow if you follow a few basic guidelines. The main requirement for dahlias is at least six hours of sun, as well as fertile, moist, well-drained soil. You can plant the bulbs out in spring, after the soil has warmed up, or you can start them inside and plant them outside later. Dahlias are not hardy in all areas. Where they are not hardy, dig them up at the end of the growing season and store them in a frost-free place until next spring.

Dahlias in a cutting garden

The bulbs are a type of tuberous root that usually has several growing points. To get an early start on spring growth, place the tubers into pots or trays of potting compost, and then water them. Give them some bottom heat above a radiator or on a heat mat. As soon as the sprouts appear above the soil, place them under grow lights or on a bright windowsill.

If you are planting them directly into the garden, dig a six-inch-deep hole in soil that is amended with plenty of organic-rich compost. Lay the tuber flat into the bottom of the hole, or if it is still a large clump of root tubers, stand it upright in a deeper hole. If you see any shoots emerging, place them facing up. Cover them with an inch or two of rich soil. Water them if your soil is dry. Do not water them again until you see the shoots above the soil. As the stems grow, gently backfill the hole with a good rich soil and compost mix until the soil is level. If you want to encourage the plant to become branched and to produce more blooms, you can pinch out the growing point when the plant has reached three or four sets of leaves or is about 8 to 12 inches high.

Tall dahlias need a strong stake that is hammered into the ground next to the dahlia tuber. Ideally, this is done during the planting process, so that you do not spike into the tuber. The flower bed does look a bit naked for a while, with only stakes and no plants. The plants eventually grow, and then you will be grateful for the stakes. You can use painted or unpainted poles or sturdy bamboo. If they are planted next to a fence, you may be able to tie them to that. Medium-height dahlias can be staked using a tomato cage, or by inserting three or four bamboo canes around the plant. Short dahlias may not need staking.

Later in the season, you will need to tie the dahlia plants to the stakes as they grow. Doing this regularly is better than coming out after a few weeks and realizing that the plants have fallen over. Little and often is the key to management of tall dahlias. Dahlias have hollow stems that resemble a pipe, so if you do not stake, they tend to blow over and crack in summer thunderstorms.

The taller the dahlia stems, the more efficiently you will need to stake and tie in your dahlias. Use biodegradable twine so that the plant and the twine can all be composted. Tie the twine around the first cane, and then loop around each cane, and so on, until you are back to the first cane. When tying up a really full plant, use five canes. Begin at the one cane and then go across the middle, loop around the cane and back across to another cane, until returning to the beginning to make a star-like pattern. As you tie dahlias in, you will also see whether they need deadheading, watering, or fertilizing. Dahlias are not no-maintenance plants.

Dahlias put on a lot of growth very quickly, so they need warmth and regular watering to grow to their full potential. They also require a rich deep soil that has plenty of nutrients. If your soil is poor, you may need to add fertilizer and mulch with extra compost. Throughout the summer into the fall, deadhead the flowers regularly to promote more blooms. Continue to tie in the stems of tall dahlias to ensure that they stand upright. Water regularly and feed with organic bloom-boosting fertilizer that has low nitrogen. Use it at a weak strength, but regularly.

Dahlia flowers are wonderful for cutting. Look in bulb catalogs for ones that say they are great for this purpose. Choose medium to tall plants that produce lots of flowers on long, strong stems.

When deciding which flowers to cut, choose ones that are open or nearly fully opened. Ideally, they will have fresh, crisp, hydrated petals with no brown marks on the back. Tightly budded blooms do not continue to open properly. Keep on cutting flowers off the plant because that stimulates the production of more buds and flowers.

Deadheading the spent dahlia flowers every day or so also encourages the plants to produce new blooms. Old flowers look very similar to the unopened buds. Look carefully as you cut so that you do not confuse the two. The new buds have a flat top, and the finished flowers come to a point and may have a few remnant petals hanging out of the tip. The plant will continue to bloom until frost, but growth slows down as the days get cooler and shorter.

If you grow dahlias in containers, they can be integrated into large pots and tubs with other summer flowers that need full sun. Make sure that the pot has stones at the bottom to counterbalance the top growth. Put in a stake to support each dahlia or grow them against a fence or decking posts to which you can tie the dahlias. Remember that dahlias are thirsty plants, so put them in with other plants that need lots of water and plant them where you can easily reach a water source.

In cold climates, the tubers will need to be dug up at the end of the growing season. We wait for a frost to blacken the tops of the dahlia plants, and then wait about another week to allow the dahlias time to send as much energy as possible from the foliage down to the tubers. In warm climates, where dahlias are hardy, they can overwinter in the ground. Some gardeners still like to lift them and store them inside for the cold months, due to the potential rotting that can occur when winters are wet. Digging up the tubers also allows you

to divide them and make more plants. Another reason for digging up dahlias is you can use that same piece of ground to plant fall-planted bulbs which you are treating as temporary inhabitants, like showy tulips.

After the waiting week, cut the dahlia stems away from their stake and chop them off at about a foot tall, leaving enough stem to use as a handle. You can always trim stems shorter later. Begin digging well away from the base of the plant. Dig around the plant in a big circle because, until you start digging, you do not know where the tubers have grown. You do not want to cut into them. When you are all the way around, begin gently levering the tubers out of the ground.

You may be surprised at the size of the tuber clump. Big plants tend to make the largest clusters, whereas some of the single-flowered dahlias and petite ones make small ones. After digging, turn the tubers upside down to allow water to drain out of the hollow stems. To increase your plants for next year, the tubers can be divided in autumn or spring. If you want to divide them, each section of tuber needs a part of the stem to start growing. Do not rip the small parts of the tuber away from the stem, but divide them very carefully with a sharp, disinfected knife. Wear gloves and cut away from yourself.

Design Ideas

Dahlias are cheerful flowers that are great to include in the garden for late summer and autumn bloom. The choice of dahlia types is wide. The type you choose will influence how they are used in the garden.

Dahlia plants have been bred to be more compact, have strong stems, and to produce

Dahlia 'White Aster' is a lovely historic dahlia from 1879. // *Dahlia* 'Knockout' // *Dahlia* 'Happy Single First Love'

more flowers. The breeding of dahlias has vastly increased the diversification of flower shape, color, and size. There really is a dahlia for every garden. Short dahlias, including some seed-grown ones, can be used to line a path, planted at the front of a bed, or popped in as annual bedding plants. Dahlias can be grown as part of a cutting area or in a vegetable-growing bed. Grow dahlias with plants that need the same growing conditions. They combine well with tomatoes and other sun-loving vegetables, as well as herbs like basil.

The medium and tall ones can be integrated into a mixed flower bed. Just make sure that you have room to access the plants to water, tie-up, and deadhead. If necessary, add a couple of stepping stones so you can reach them. Use dahlias at the middle to back of the bed, and grow them with other late bloomers like gladioli, galtonias, salvias (*Salvia*), flowering tobacco (*Nicotiana*), cosmos (*Cosmos*), and zinnias (*Zinnia*).

If you have never tried dahlias in your garden, I would highly suggest adding a few this coming summer. Once you are hooked, there might be more next year.

Species and Cultivars

Dahlias are divided into groups, according to the shape and size of their blooms. While it is not necessary to know the classification types to grow and appreciate dahlias, it is fun to be able to see the possible ones that you could grow.

When making your selections, look at the height of the plant and also the color of the foliage. Low-growing dahlias are not in a category by themselves but have small flowers on short plants that are good for bedding, edging, and containers. Dark-foliaged plants look stunning in mixed flower beds. The classic dark-leaved dahlia is 'Bishop of Llandaff' with red flowers. There are also variable, seed-grown offspring

CLOCKWISE FROM LEFT: *Dahlia* 'Sandia Brocade' // *Dahlia* 'Lemon Sherbet' // *Dahlia* 'Star Child' // *Dahlia* 'Verrone's Obsidian'

from this bishop that are called 'Bishop's Children' that have various hot-colored flowers. 'Knockout' and 'Yellow Hammer' are good choices for yellow flowers and dark leaves.

Single dahlias have daisy-shaped, open blooms that are pollinator friendly. They have a lovely flat face with 8 to 10 slightly overlapping rounded-tipped petals. Single dahlias are not the best as cutting flowers because their petals drop off quickly once they are picked. They bloom early in the dahlia season. Singles look good growing in containers or as a mixed plant combination in a flower bed. Short single dahlias are great when planted as a row, lining the front of the border. There are many to choose from that vary in flower colors. 'Happy Single First Love' and others in the Happy Single Series are all great in the garden.

Anemone-flowered dahlias have an outer ring or rings of petals that are similar to singles, but with an additional pincushion-like grouping of central small petals. These have a different look, so are great for mixed dahlia bunches. 'Platinum Blonde' is a light creamy yellow, 'Totally Tangerine' is a popular soft orange, and 'Sandia Brocade' is a good coral yellow.

Collarette dahlias are similar to anemone-type dahlias. They have an outer ring of petals with smaller frilly inner petals in a ring around an open yellow center. Some of the best include pink-and-white 'Bumble Rumble', orange 'EZ Duzzit', and light yellow 'Lemon Sherbet', which is part of the Sweet Candy Series.

Orchid- and star-shaped dahlias have an open center surrounded by windmill-shaped flowers. The petals are narrow and rolled inward at the end. There are some double orchid-shaped dahlias. 'Honka' is a popular yellow, 'Star Child' is a full-looking white with flat petals, and 'Verrone's Obsidian' is a very dark mahogany, bordering on black.

CLOCKWISE FROM LEFT: *Dahlia* 'Brown Sugar' // *Dahlia* 'Tahoma April' // *Dahlia* 'Ginger Willo' // *Dahlia* 'Moor Place' // *Dahlia* 'Fusion'

Ball dahlias have fully double flowers with pleasing densely packed balls of petals. Each petal is rolled up almost into a tube. Bloom size is three to four inches across. These plants usually make plentiful flowers for cutting. I love their symmetrical shape, both in the garden and in a vase. 'L'Ancresse' is a fabulous white, 'Polventon Supreme' and 'Nettie' have yellow flowers, and 'Brown Sugar' has rich brownish coloring. 'Tahoma April' is a lovely dark pink that is fabulous for cutting.

Pompon dahlias are miniature balls, usually less than two inches across. They are perfect for small-scale flower arrangements. The plants are usually short and slender. Some favorites include pink-flowered 'Betty Anne', two-tone orange 'Ginger Willo', dark maroon 'Moor Place', and 'Small World' for a diminutive white.

Cactus dahlias are fully double flowers with rolled petals that are pointed at the ends. The petals are arranged in a regular pattern. Their distinctive shape looks good in a mixed flower bed or in a vase. 'Tanjoh' has a white center and pinky purple tips to the petals; 'Fusion' is a rich orange with yellow at the center of the flower; and 'Tutti Frutti' has pink petals that are also yellow at the center.

CLOCKWISE FROM LEFT: *Dahlia* 'Poppers' // *Dahlia* 'Jane' // *Dahlia* 'Skipley Spot' // *Dahlia* 'Kelvin Floodlight' // *Dahlia* 'Mikayla Miranda'

Semi-cactus dahlias are similar to the cactus shape, but they have a flat base to each petal with the tips being rolled or quilled. 'Hokey Pokey' and 'Miss Rose Fletcher' have pink flowers. 'Poppers' is an amazing red and yellow.

Incurved cactus dahlias have rolled petals that bend in toward the center of the flower for a very distinctive look. I love the bright orange 'Bed Head'—in my garden it grows to be an eight-foot-tall plant. Also look for the pinky purple flowers of 'Jane'. These plants are distinctive and different in the flower garden or in arrangements.

Formal decorative dahlias have flat petals held in a symmetrical arrangement. There are many dahlias to try in this group of popular garden plants. They vary in size but some of these get to be the size of dinner plates. These are fabulous for large-scale floral designs. Look for the rich maroon-purple 'Tootles', the reliable rich purple 'Thomas Edison', the large fresh yellow blooms of 'Kelvin Floodlight', and the amazing red-and-white-patterned 'Skipley Spot'.

Dahlia 'Drama Queen' // *Dahlia* 'Onesta' // *Dahlia* 'Aphrodite'

Informal decorative dahlias may have flat or slightly rolled petals that have a rather disorganized pattern of petals. This is another group that provides some great choices for both cutting and border plants. One of my favorites in this group that produces many blooms for me is 'Mikayla Miranda', which is white and pale lavender-pink. The petals of fuchsia-pink 'Drama Queen' have an attractive drooping look at the tips. 'Gitts Attention' is a full white flower with slight notching at the ends of the petals. One of the strongest and most forgiving dahlias that I have grown is 'Bahama Mama'. It is a good multiplier and produces plentiful pink-and-yellow flowers.

Waterlily-shaped dahlias have rings of petals held in a spiral on a flattened base. The petals have an upward curl to their edges. The center is often closed. These plants are short in stature, so they integrate well into mixed flower beds and are good in containers. Choose 'Lemonade' for a good mid-yellow, 'Patricia Ann's Sunset' for a great red-orange, or the pink-flowered 'Onesta'.

Laciniated or fimbriated dahlias have a distinctive split at the tips of the petals to give a feathery look to the whole flower. 'Citron du Cap' is a great garden plant and excellent for cutting. It has a yellowy center that blends to pink at the petal tips. 'Show 'n' Tell' is a large red flower with a little yellow at the tips. 'Aphrodite' is a lovely fimbriated dahlia with creamy yellow in the center, ending at dark pink, fringed petal tips.

Eucomis

eucomis, pineapple lily

Full sun to part sun // Moist, well-drained soil // Zones 7–10 // Height 24–36 in. //
Season of interest—summer into fall

About This Bulb

These bulbs are a fun addition to a mid to late summer display. This bulb is called a pineapple lily, although it is neither a lily nor a pineapple. Its common name comes from its growth habit, which resembles a pineapple plant, with its clump of wide, glossy, strap-like leaves and a cylindrical spike of flowers that emerges from the center. Pineapple lily leaves are about two or three feet long, depending on the type. Some species have wavy leaf margins. Foliage color can be bright green or burgundy-maroon. Speckles are sometimes present on the leaves.

The strong, upright flower stems emerge from the center of the foliage, in a similar way to a developing young pineapple. Each individual flower is small, but they are densely packed up the stem. A tiny tuftlet of foliage, which is actually composed of leaf-like bracts, sits on top of the flower. The flowers begin opening at the bottom of the flowering stem, and sequentially open toward the top. Flower colors are green, yellow-green, cream, white, pink, or burgundy. Each flower looks like a little star with six petals. Eucomis makes a good garden plant where it is hardy, providing a strong foliage statement, topped with a dramatic pineapple-like flower spike in summer.

The first pineapple lilies that I grew were in a pot. I had heard of them, but never seen one. The flower stalk was a great surprise when it emerged. For many gardeners, growing in containers is the best choice. These bulbs require good soil drainage and hot summer weather to grow well.

Planting and Maintenance

Pineapple lilies need a position in at least part sun. If you garden in a warm climate, plant them in bright shade, in a place where the bulbs do not get wet during winter. In climates with hot summers, protect them from strong afternoon sun, and water regularly. Plant the bulbs six to eight inches deep, depending on your soil drainage. Plant deeply in a sandy soil, and closer to the surface in clay soil. Eucomis need plenty of water when they are growing.

Where they are hardy, you can overwinter them in the ground. If they are borderline hardy, choose a protected spot and cover the ground with evergreen branches, salt hay, or pine straw. They are not evergreen so they will lose their leaves in winter. Bulbs may need dividing after three or

Eucomis bicolor leaves // *Eucomis pallidiflora* subsp. *pole-evansii* with pollinated flowers // *Eucomis comosa* // *Eucomis comosa* 'Sparkling Burgundy'

four years; however, it's generally best to leave pineapple lilies in the same position unless you see that they are not blooming well.

If your climate is not ideal for growing pineapple lilies, it is easier to grow them in containers. The soil should be moisture-retentive with high organic matter, but also extremely well-drained, especially during the resting period in winter. Try adding grit or gravel to an organic-rich planting mix. In a container, plant the bulb so that the neck is showing. This is a trick to prevent rotting. If you have space inside in a garage, or cool greenhouse, you can start watering the pineapple lily bulbs in spring before your last frost date. It takes a few weeks to begin growth, so this way, you are getting an early start to the growing season. The bulbs can be planted outside once the nighttime temperatures warm up

to about 60 degrees F. Sit the container on pot feet or tiles to allow water to drain out of the bottom of the pot.

During the dormant time, in the cold months, keep the bulbs relatively dry. As the leaves wither away in the autumn, you can carefully remove them. In spring, bring the plant into a warm area and start watering again. If your pineapple lilies are planted in a large pot that cannot come inside, or to save storage space, you can take the bulbs out of the pot and keep them in a cool, dry place until spring.

Design Ideas

As a container-grown plant, pineapple lily is versatile. Place it in the flower bed, grow it alone in a pot, or combine it with other summer plants with finer textures. I really like to grow pineapple lily in a pot by itself, especially in an elevated place like on a wall or table. Try it in an urn-shaped pot to mirror the shape of the flower and fruit. However, pineapple lilies also combine well with other summer bulbs, tropical plants, annuals, or perennials.

Eucomis has such an unusual flower shape and broad elongated leaves that it will stand out above its plant neighbors, especially the very tall ones like *E. pallidiflora* subsp. *pole-evansii*.

Species and Cultivars

Eucomis bicolor lives up to its name with stems and leaves that are green with dark-colored spots. Flowers are creamy white with a green crown of leaf-like bracts on top. It is not a good cut flower because it has an unpleasant odor. It is about two feet tall at blooming. *E. bicolor* 'Alba' has off-white flowers. It is hardy in zones 8 to 10.

Eucomis pallidiflora subsp. pole-evansii grows to a giant three to six feet. It blooms in late summer and early autumn with greenish white flowers that look green when pollinated.

Eucomis comosa is one of the best for in-ground and container culture. The strap-shaped, pointed leaves form a loose rosette. Foliage color is green or green-washed with mahogany. The sweetly scented flowers are variable. Some are pinkish white, while others are burgundy-tinged with a dark-colored ovary. A small tuft of foliage appears at the top of the flower. This species grows up to three feet tall. It blooms in late summer, is good for pollinators, and makes an unusual cut flower. 'Sparkling Burgundy' is 16 to 20 inches tall with purple-bronze foliage, dark stems, and purple buds that open to pink, beginning from the bottom to the top of the flower.

About This Bulb

Galtonia is a genus of summer-blooming bulbs that is rather underused. They make a lovely addition to a bed or container garden. Flower spikes holding dozens of dangling, bell-shaped, fragrant, white or green flowers are suspended from two- to four-foot stems. Depending on planting time, the flowers bloom in mid to late summer. The strap-like foliage is green, and the plant has a vase-shaped appearance overall. It is an easy-to-grow garden plant that looks rather like a tall, thin hyacinth, giving rise to its common name. Galtonias can be used as cut flowers.

Planting and Maintenance

Plant the large galtonia bulbs about six to eight inches deep in a bed or in a pot with plenty of room for good root development. They need a warm site with good drainage, but with consistent moisture. In cold climates, they should be planted outside in spring, once the soil has warmed. Where they are frost hardy, they can be planted in the autumn. As additional protection, cover the soil surface with a thick layer of lightweight mulch in the autumn. Some of this can be removed in spring as the soil warms up. Watch out for slug and snail damage when the plant is young. Apart from this, they are relatively pest resistant.

If growing galtonias in a container, choose a large one. Plant them a couple of inches deep, water them in, and place the pots in a sunny location. Bring the pots inside for the winter and keep them dry and cool, but not cold, for winter dormancy. I like to plant them in raised beds where the soil can be easily amended.

Design Ideas

The tall stems of galtonia look good planted to rise above surrounding summer annuals and perennials. They work well in cutting gardens, fragrance gardens, cottage gardens, or mixed flower beds. Arrange the large bulbs in groups, or clusters, for maximum impact, because the stems are narrow. Try them with perennials like catmint, or annuals such as marigolds (*Tagetes*), zinnias (*Zinnia*), or cosmos (*Cosmos*). My middle daughter has them growing in among her peonies, where they add some summer interest after the peony flowers have finished. Summer hyacinths also associate well with other bulbs like crocosmias, gladioli, lilies, and agapanthus.

Galtonia candicans // Galtonia viridiflora

Species and Cultivars

Galtonia candicans, summer hyacinth, is the most commonly grown species in this genus. The white, bell-shaped flowers are delightfully scented. This plant is two to four feet tall.

Galtonia viridiflora, green-flowered galtonia, is a lesser-known species that has lightly scented greenish cream flowers borne on a two- to three-foot stem. The flowers are pendulous and clustered with a wider appearance than *G. candicans*. The flowers open in succession from the bottom to the top rather than all at once. To grow well, this bulb requires fertile, well-drained soil that stays reliably moist all summer. Where it is hardy in the ground, it may increase in height for the first few years while it is getting established.

Gladiolus

gladiolus

Full sun // Moist, well-drained soil // Zones 7–10 // Height 24–48 in. // Season of interest—late spring through fall

About This Bulb

Gladioli provide perfect tall flower spikes for the summer garden. They are easy-to-grow bulbs. In cold winter areas, plant the bulbs in spring. In warm climates, plant them in spring or fall. One or more gladiolus flower spikes and a spray of bright green, sword-shaped leaves grows from each bulb, which is technically a corm. Big bulbs yield more flowers, and each bloom is larger in size. It is well worth getting top-sized ones from a reliable source for this reason. The flower stalk gradually elongates from the middle of the foliage. The blooms open sequentially from the bottom to the top of the stem, to give an extended period of interest in the garden.

Gladiolus flowers are available in an extensive range of colors from cool whites, creams, and light pinks, through mid-tone purples, peaches, and bold brilliant colors such as bright fuchsia, red, orange, and yellow. Flowers may have a different color in the center, a wash of several hues, or edges with a contrasting outline. Some gladioli are simple, slender, and delicate, while others are large, complex, and blowsy, maybe even with ruffled petal edges. Each individual flower is trumpet- or funnel-shaped. Flower spikes have all the flowers on one side only, and they face to the sunny side of your garden. A few gladiolus flowers have the added delight of fragrance, often only in the evening. Choose your perfect gladiolus by looking at the flower color, size, shape, and number of blooms per stem, whether or not it has fragrance, and the height of the flower stalk. They are deer and rabbit resistant.

Gladioli make a dramatic flower both in the garden and cut in a vase. They bring excitement to a mixed flower bed, a cutting garden, the edge of a vegetable garden, or a large container. You may see the word "gladioli" shortened to "glads."

Planting and Maintenance

Plant bulbs in a sunny spot in rich, well-drained soil. Wait to plant them outside until the soil has warmed up after the last frost. Dig a hole and put sand or grit in the bottom as a little cushion for the bulbs. If you are planting them in the ground, you can plant a large corm six to eight inches down and smaller ones at about four to six inches. Planting the corms deeply stabilizes the tall plant in the soil and helps them to overwinter in mild climates. Use a mixture of grit and organic-rich compost, mixed with your regular soil, to partially fill in the hole after planting. Continue to backfill the hole as the plant sprouts up, ending up with the soil being flush with the surface.

If you are planting rows in a cutting garden or vegetable garden, dig a narrow trench and lay the bulbs on gravel at the bottom, and again, gradually fill in as they grow. The stems may still need staking but the deeply planted corms tend to have some support from the weight of the soil around their stems. If needed, add a bamboo cane to hold them upright. Another trick is to plant them against a fence for support. I grow my glads in a raised bed because it is easy to dig down deeply into the soil to plant them.

Water the soil after planting and keep them well watered during active growth. When planted in organic soil, gladioli are unlikely to need fertilizing. If you are digging up the corms to save them over winter, it is a good idea to fertilize them after flowering. Use an organic seaweed or fish emulsion.

If you plant dozens at once, you will have every flower in bloom at the same time, which may be more than you need for one vase. To extend the bloom time of gladioli in your garden, stagger the plantings. Plant some bulbs the first week after the soil warms. Then plant a few more every few weeks until midsummer. This will produce a succession of blooms through the summer and into early fall. A bulb takes a minimum of two to three months from planting to bloom, so do not plant your last batch too late.

In cold winter climates, the bulbs will need digging up in the autumn and storing in a cool, frost-free place. Dig up the whole plant when the foliage begins to turn yellow or brown. Place them into a cool situation in a crate or cardboard box to dry out and cure. Alternatively, you can

tie a bundle together and hang them up to dry or place them into a paper bag. When the bulbs and tops have dried, you can cut the stems off, but leave an inch or two attached.

In spring, before you plant them back out into the garden, remove the old, shriveled bulbs at the base and the papery covering by rubbing them with the back of your thumb. If you want, you can keep the little baby cormlets or cormels that will not flower this year. They can be saved and potted up. They will flower after a couple of years.

Design Ideas

Gladioli are great bulbs for interplanting in a mixed border or in a bed of warm-season annuals. Small groups of bulbs can be easily popped into the soil between emerging perennials and annuals in spring. They work well when planted in the middle of the bed, so that as the foliage dies down, it is hidden by surrounding plants. You can repeat the same gladiolus in waves throughout the planting to unify the area.

The traditional place to grow gladioli is in a row in a cutting garden. Another option is to plant some at the edge of a vegetable garden, where they can be easily enjoyed in place or harvested for cutting. Gladioli make a dramatic, vertical cut flower. Pick them when about two thirds of the flowers are in bud, and some of them will continue to open in the vase.

Gladiolus communis subsp. byzantinus // Gladiolus murielae // Gladiolus 'Plum Tart' // A lovely red gladiolus

Species and Cultivars

Gladiolus communis subsp. byzantinus

has a slender, upright stem measuring two to three feet, on which bright magenta, funnel-shaped flowers are arrayed. They bloom earlier than most gladioli. Where they are hardy, they may be in bloom by mid to late spring. They are hardy in zones 6 to 10, but outside these limits, they can be treated like a tender bulb, dug up and replanted the next growing year. They need a free-draining soil to do really well. They are perfect bulbs for planting in a mixed flower bed among perennials and annuals. The magenta color of this gladiolus is a favorite of mine.

Gladiolus murielae, acidanthera, has large, white, triangular-shaped flowers with a purple-burgundy blotch in the center. The flowers are held toward the ends of arching stems, above spiky green leaves, and open in sequence over a few weeks. The pallid blooms shine out at dusk on an early autumn stroll around the garden. They are fabulously fragrant at that time of night.

They are easily grown in a flower bed or container. Plant the bulbs two to three inches deep. Where they are not hardy, dig them and store in a frost-free place for winter. If they are hardy, dig them up every few years to divide the clump, and add some fresh organic matter into the hole. Alternatively, add an annual top-dressing of organic-rich material to the top

Gladiolus 'Green Star' // Rich purple gladiolus // *Gladiolus* 'Cantate' // *Gladiolus primulinus* 'Las Vegas'

of the soil after flowering. Full sun is best in most climates. Afternoon shade is needed in hot areas. Acidantheras need moist but well-drained soils containing plenty of organic matter.

Hybrid gladioli are divided into two main groups, depending on size of stem and bloom.

Large-flowered hybrid gladioli are what most gardeners think of when you say you are planting glads. They are the darling of the cut flower trade for their dramatic long stems that hold up well in water. The colors are bright and cheerful, bordering on garish. Red, orange, yellow, white, pink, magenta, and fuchsia are popular colors. Some flowers have a contrasting color that looks like an eye.

Small-flowered hybrid gladioli are similar in look but petite and dainty. They grow two to three feet tall, and their blooms are delicate, often with a slightly triangular flower shape. The color range is wide with some bicolors and tricolors. Look for gladioli labeled *nanus* or *primulinus*. These little beauties look lovely in containers or at the front of flower beds in combination with low-growing perennials.

Gloriosa

gloriosa lily

Full sun // Rich, well-drained soil // Zones 8–10 // Height 36–72 in. // Season of interest—summer into fall

About This Bulb

Gloriosa lilies produce vines from their bulbs. During the course of the summer, they grow three- to six-foot-long stems that hold shiny, elongated leaves ending in tendrils. These tendrils are used by the plant to grab onto its supports and climb up toward the sun. They need to be supported or planted next to a fence or other plants that they can hold onto and scramble into. These flowers are showstoppers. The foliage disappears into the background, but the bright blooms catch your attention as you pass by. Gloriosa lily flowers live up to their name with their beauty. They are not commonly grown, so when gardeners see them, they admire them.

The delicate four-inch flowers have twisted petals that turn back on themselves. The flowers have prominent stamens that jut outward from the flower and are part of the decorative appeal. The overall effect is of a graceful butterfly-like bloom that appears to float and dangle among the foliage. I have seen gloriosa lilies used in cut flower arrangements. They are as stunning dangling from a vase as they are in a garden. I have not tried this personally because it seemed a shame to cut my few blooms. They are pest resistant.

Planting and Maintenance

Plant your gloriosa lily bulbs horizontally, about three inches deep, when the soil has warmed in late spring. They are fragile. Handle them with care so that you do not break them. If you are planting them into a container, add extra grit to the planting mix to increase drainage. If your summer is short, start the tubers a few weeks before your last frost date in a warm place inside. The bulbs take a while to break dormancy, and the plants need a long growing season with hot weather to produce flowers. Flowers are produced over an extended period, as long as the air and soil remain warm.

Outdoors, choose a planting spot next to a fence or other supporting structure, so that the vine can climb it. An ideal situation is where the roots are cool and shady, and the top growth gets plenty of sun for the leaves. Plant the bulbs on the shady side of surrounding plants. Another trick is to cover the edge of the bulb with a flat stone once the shoot has emerged. The stone adds shade and keeps vital moisture locked into the soil. If the gloriosa lily is grown in a plant-packed flower bed or container, it will scramble through its neighbors in a way that is similar to its natural

Gloriosa superba 'Rothschildiana'

Gloriosa bulbs may need protection from cold and wet in winter, even where they are supposedly hardy. In the autumn, add a couple of inches of an open and airy mulch like salt hay, or pine straw, to the soil surface, for added protection. In areas with cold winters, dig and store the bulbs in an area that is kept above 50 degrees F. If you are growing bulbs in a pot during the summer, leave them in the same container over winter. Then you do not have to handle and potentially damage the fragile bulbs. All parts of this plant are poisonous so handle with gloves.

Design Ideas

Gloriosa lilies are ideal plants to grow up and through surrounding summer plants. They will need either plants or supports to grow into, to show off their incredible flowers. A gardening friend of mine used glory lilies as the centerpiece of a huge container that had a tripod structure in the middle. The vines climbed up the support and were surrounded by other wonderful sun-loving plants like cannas, salvias (*Salvia*), and sweet potato vine (*Ipomea*). Grow them either in the ground or in a large pot.

Species and Cultivars

Gloriosa superba 'Rothschildiana'
must be a great plant to grow, with a name that includes both "superb" and "glory." It is a good grower that has bright red flowers with yellow in the center and a yellow margin to their petal edges. *G. superba* 'Citrina' has the same flower shape but is predominantly yellow.

habitat. I love the effect of the gloriosa growing up and through other plants.

The bulbs and stems can easily be damaged, so put stakes around them in the ground at the time of planting. If you stake at a later time, you risk spearing the bulb and damaging it. Choose wires or sticks that are narrow in diameter to enable the tendrils to attach easily. Where this bulb is hardy, and can develop in long hot summers, the vines can reach six feet. In gardens with short summer seasons, or where there is a restricted root volume, like in a container, the ultimate height may be three feet.

Hedychium

hedychium, ginger, ginger lily

Full sun to part shade // Rich, moist soil // Zones 7–11 // Height 36–96 in. // Season of interest—summer into fall

About This Bulb

Ginger lilies, or *Hedychium*, have a sturdy and strong presence in the garden. Their large, lance-shaped, glossy leaves are held alternately up their strong stems. The tube- or trumpet-shaped individual flowers are clustered into elongated spikes or groupings at the top of the tall stems in summer. Flower colors are orangey melon, orange, red-orange, yellow, or creamy white. Some have intensely fragrant flowers. They can be grown in large containers or in the ground.

Planting and Maintenance

Plant the bulbs, that are rhizomes, close to the soil surface and give them copious amounts of water while they are in active growth. They are large plants that grow rapidly, so they also require a rich soil that has been amended with your best compost. To keep the bulbs moist and provide slow-release nutrients, top-dress with more organic matter as the season progresses. If you feel that your plants are lacking vigor, then feed them with a good potassium-based multipurpose fertilizer. Some gingers are understory plants in their natural habitats and seem to thrive with part shade. In hot climates, choose a place with sun in the morning, and shade during the hot summer afternoons. In temperate climates, give gingers a position with at least half a day of sun.

Gingers vary in hardiness according to which species and cultivar you choose. In gardens with mild winters, they can stay in the ground with a fall addition of mulch. Where they are borderline hardy, try planting them on the sunny side of your house. If you are digging them up for the winter, wait for the leaves to be nipped by frost, and then dig around the root zone, giving them a wide berth, so as not to damage them.

If you are growing your ginger in a container, bring the whole plant inside for the winter and ease off on watering to allow the plant to go dormant. If you have them in a dry place, they may need occasional watering in the cold months. When your plant becomes too large for the pot, tip the whole root ball out of the container, and divide it.

Design Ideas

When designing with gingers, use them as a bold focal point, a repeated element down a flower bed, or as a tall background to a bed. If they are hardy, they will form impressive stands that can

Hedychium coronarium // Hedychium densiflorum // Hedychium 'Tara'

create a summer screen. Combine them with other bulbs that grow in the same conditions, such as cannas, colocasias, and alocasias. Leave enough room between the bulbs so that they look good together visually and can each get enough water, nutrients, and sunshine.

Grow hedychiums in large containers that suit the area where they will be growing. Put them in position and then plant them up because they will be difficult to move once the plants are growing. There are dozens of tropical species of gingers, so look for ones that are appropriate in size and color for your garden.

Species and Cultivars

Hedychium coccineum, scarlet ginger, has long, narrow leaves with a white midrib and is a structurally strong plant. It is lightly fragrant, heavily blooming, and hardy in zones 8 to 10. Some good cultivars include 'Orange Brush', 'Devon Cream', and the pale yellow 'Tresco'.

Hedychium coronarium, white butterfly ginger, has white flowers with a yellow center. It grows three to six feet tall. White butterfly ginger has an alluring scent that carries on the wind and draws both pollinators and people to find the source. When we grew it in a public garden where I worked, it was the hit of the summer. 'Maximum' is a large-flowered cultivar, and *H. coronarium* var. *chrysoleucum* has a light yellow flower center. They are hardy in zones 7 to 11.

Hedychium densiflorum, orange butterfly ginger, is hardy in zones 8 to 10, and grows to about three feet. It has bold orange flowers packed together into a bottlebrush shape. This ginger is best for temperate climates. In hot areas, provide shade, especially in the late afternoon. 'Assam Orange' is one of the most popular with great bright orange flowers. 'Sorung' has peachy apricot flowers. 'Stephen' has large glossy leaves and yellow-orange flowers with deeper orange stamens that appear in late summer.

The hybrid ginger *Hedychium* 'Tara' has orange flowers and grows from four to six feet with fragrant dense groupings of flowers in late summer.

Hesperantha

hesperantha, schizostylis, crimson flag

Full sun // Rich, moist soil // Zones 7–10 // Height 18–24 in. // Season of interest—late summer into fall

About This Bulb

Hesperantha coccinea blooms in autumn with groupings of slightly cup-shaped flowers composed of six petals. Flower colors range from bright scarlet red to coral to light pink. It has fans of evergreen foliage. These bulbs grow best in mild winter climates. An older name for this plant is *Schizostylis*—you might still see it for sale under that name.

Planting and Maintenance

Plant the bulbs two to three inches deep in an area with rich soil and full sun. Water these bulbs well in the summer to promote good fall flowering, and feed if necessary. If they are borderline hardy, mulch them heavily before winter. If growing them in containers, bring them inside for the winter and leave them undisturbed in their pots. Keep them moist, but not wet, at temperatures above freezing, and take them out again after spring weather is consistently frost-free.

Design Ideas

Hesperantha looks good when planted in the mid zones of a rain garden, where the soil is moist but not constantly wet. They can grow in regular garden beds, as long as the soil is rich and retains moisture. They look good coming up through groundcovers or randomly placed throughout a mixed flower bed in small clumps. Where they are not hardy, plant them in a pot that can be grown on a sunny terrace. They provide a nice autumnal flower surprise.

Species and Cultivars

Hesperantha coccinea is the only species in this genus. The cultivars are well worth seeking out for their specific colors and more profuse flowering. Look for 'Sunrise', with its large salmon-pink flowers, white 'Alba', and rich red 'Major'.

TOP: *Hesperantha coccinea*
BOTTOM: *Hesperantha coccinea* 'Sunrise'

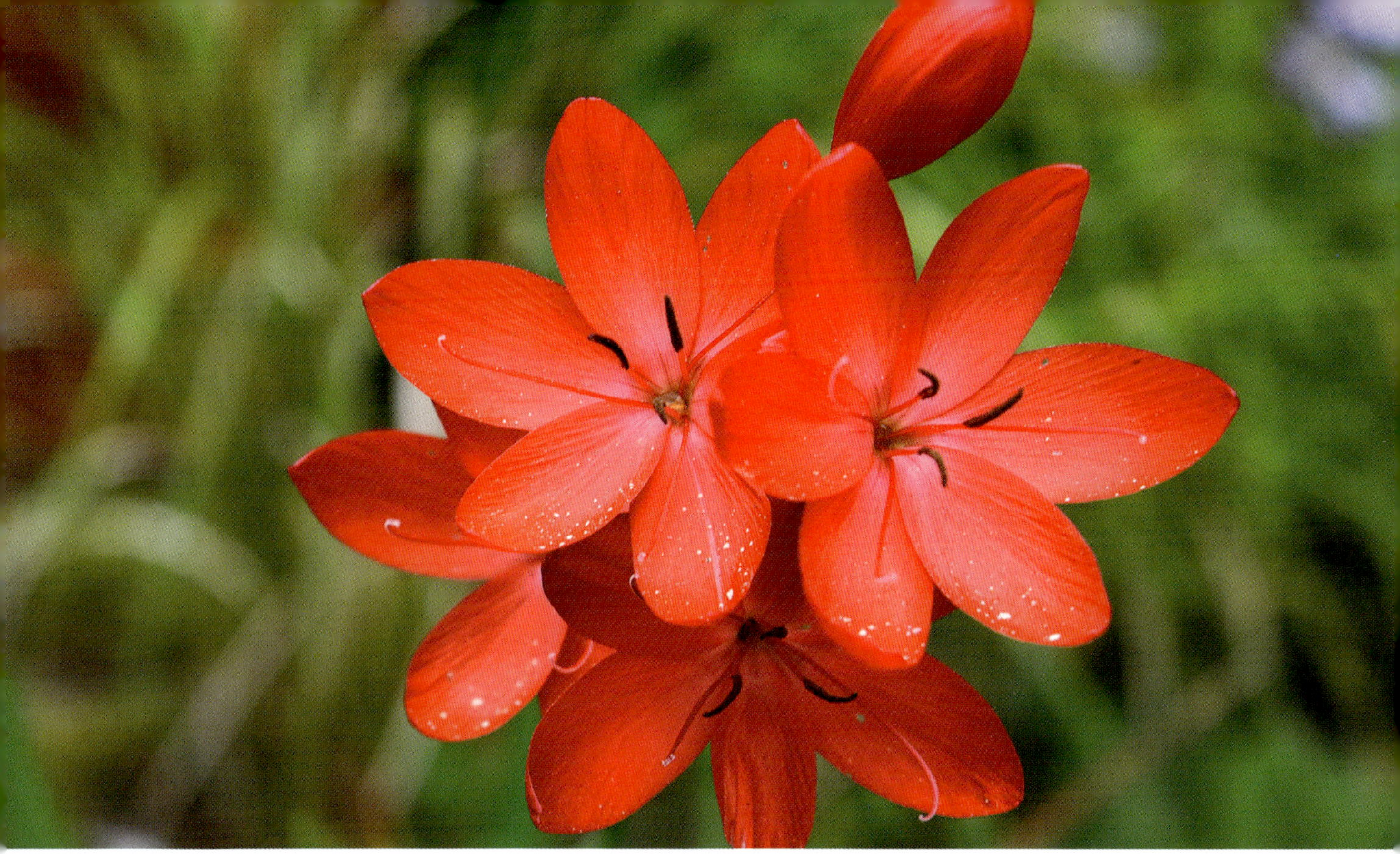

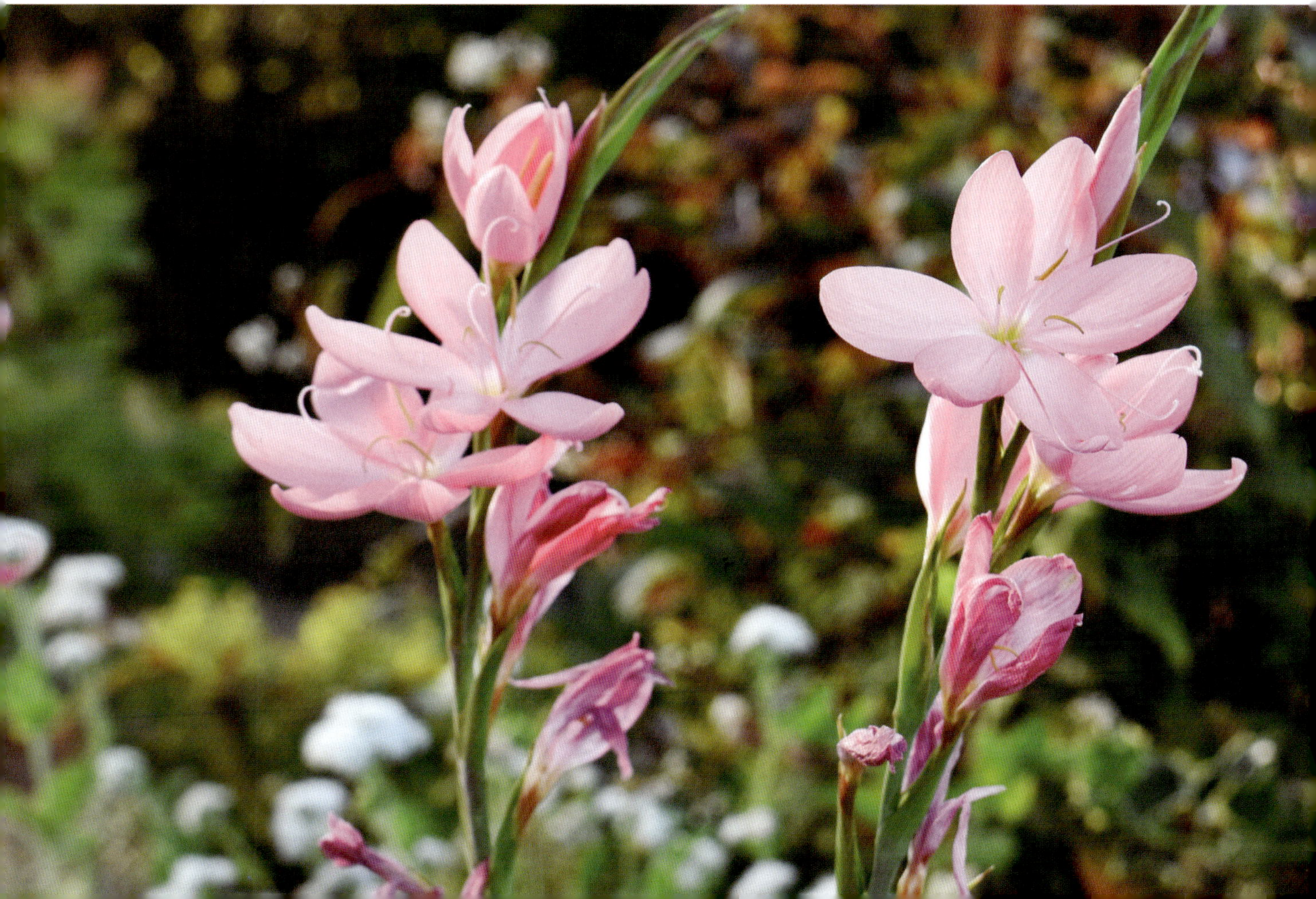

Hymenocallis

hymenocallis, ismene, spider lily

Full sun to part shade // Rich, moist, well-drained soil //
Zones 8–11 // Height 18–30 in. // Season of interest—summer

About This Bulb

Hymenocallis is a genus of plants that comes from the southeastern United States and South America. It grows best in similar climates that have mild winters and hot summers. This bulb has a weird and wonderful flower shape. The central core is a round cup that has six stamens attached to the inside. The cup is ringed with long outer petal-like structures that may have a slight twist to them. You can see how they could resemble spider legs. However, to be a spider, it would need eight petals or legs. Several other plants are also called by the common name of spider lily, so it can be a little confusing.

The flower is often a pure, crisp white, and shows up well against its strap-shaped, mid-green leaves. There are pale yellow cultivars, which have striking inner green markings. This coloration is visible when you look from the front of the flower. The foliage is held like many bulbs, as an upright rosette attached at the base of the plant. The leaves of some species are deciduous, and others are evergreen. If you are growing them in a large pot and overwintering them inside, the deciduous ones are the best choice. In areas where they are hardy, you can grow any or all of them. The lovely flower fragrance is an added benefit.

There are different species, some of which used to be called ismenes. Not all of them are commonly grown in gardens, but if you live in the native range of this plant, it would be well worth seeking out more types from a native plant nursery. In my zone 6 to 7 garden, I grow hymenocallis in a pot. Where it is hardy, hymenocallis is a great weed-smothering plant that is a favorite old-fashioned, pass-along plant.

Planting and Maintenance

Hymenocallis can be grown in the ground or in containers. Wait until the soil has warmed up before planting in spring. In the ground, plant the bulbs about four or more inches deep. Plant deeply if these bulbs are borderline hardy in your climate. In a container, you can plant shallowly with the neck of the bulb exposed. The plant grows rapidly with leaves first; then long flexible flower stems with elongated buds appear. The buds burst open to reveal the spidery-looking flower. During their growing season, water these bulbs well. Add some potassium feed to the water while they are growing rapidly.

Hymenocallis ×festalis 'Zwanenburg' // Hymenocallis harrisiana

Deciduous plants are the easiest to overwinter because you can let them lose their leaves for storage. Gradually let the soil dry out as the fall temperatures drop. When the leaves turn yellow, lever them out of the ground, taking care not to damage too many roots. Turn them upside down to dry for a few days in a frost-free place. Then store in a dry cool place, but do not let them get cold. About 50 degrees F will work, but not much colder in dormancy. You can leave the bulbs in their original pot if you have space to store them that way until next spring.

Design Ideas

Where hymenocallis are hardy, they make great garden plants for a slightly wild area. When they are grown in a pot, they can be brought out to a patio or terrace when they are in bloom. The unusual flower shape and their fragrance are always topics of conversation. In areas with cool summers, they are best grown inside.

Species and Cultivars

Hymenocallis ×festalis 'Zwanenburg' is a vigorous hybrid plant. It has large, scented flowers with crenellated edges to the cups. It can bloom from late spring into summer, depending on planting time.

Hymenocallis harrisiana is from Mexico. It has very narrow outer petals that are attached to a small cup to produce an elegant flower. Where it is hardy, the plant can develop into a large clump over time.

Hymenocallis narcissiflora, ismene, has larger cups than some in this genus. The fragrant summer flowers can be tinged with green or creamy yellow. Citrus-scented *Hymenocallis* 'Sulphur Queen' has a lovely star-like pattern in the inner center of the yellow flower. These plants are hardy in zones 8 to 11.

Liatris

liatris, blazing star

Full sun // Average or moist, well-drained soil // Zones 3–9 //
Height 24–60 in. // Season of interest—summer into fall

About This Bulb

Liatris are delightful North American wildflowers that make lovely vertical accents in the summer to early fall garden. Each inflorescence is made up of hundreds of individual flowers that the pollinators love. Butterflies—as well as smaller insects, such as hoverflies—are particularly attracted to blazing stars. Flower colors are violet, purple, or creamy white. Each spike has a long bloom time because the individual flowers open in sequence from the top to the bottom. They can be used as cut flowers, either fresh or dried.

The foliage of blazing stars is attractive before flowering. The slender, lance-shaped leaves form a basal rosette that is tuft-like in appearance. Additional slim leaves are arrayed up the flower stalk.

Planting and Maintenance

Liatris are sometimes sold as bulbs or purchased as potted plants. If you buy a liatris that is already growing, plant it as you would any herbaceous perennial, making sure to keep the soil level the same as it was in the pot. If you buy bulbs, plant them a couple of inches below the soil surface.

Liatris can take hot summers, but cannot tolerate wet winters, so choose a slightly elevated planting site. Some of the blazing stars that are naturally found in dry soil make perfect plants for a gravel garden that is rarely watered. I have even seen liatris growing out of cracks between rocks. It seems to do well there because it stays moist under the rock, but the soil is well-drained. They can also be planted in containers.

When you see liatris growing in the wild, the soil looks fairly rocky. Copy this in your garden by growing liatris in soil amended with extra grit or gravel. Excess fertilization leads to floppy top growth especially with the taller species and cultivars. I much prefer to restrict organic matter and fertilizer to forgo the need for staking. Growing liatris among surrounding bushy plants helps to keep them standing upright.

After the flower stalk has finished blooming, you can cut it off at the base. If you garden for wildlife, leave the stalk on the plant. The seeds are appreciated by birds, like finches and other seed eaters. To make more plants, you can save some seed and sprinkle it around the garden, or sow it in trays of gritty compost. Seedlings take a couple of years to get to flowering size.

Liatris aspera // Liatris microcephala // Liatris pycnostachya

Design Ideas

Liatris is a useful plant because its small bulb size means that it is easily planted between surrounding perennials. It looks particularly fabulous among rudbeckias (*Rudbeckia*), echinacea (*Echinacea*), helenium (*Helenium*), and salvias like *Salvia* 'May Night'. Liatris is a great choice for slightly wild-looking areas, prairie-style plantings, summer meadows, and as a cut flower.

Species and Cultivars

Liatris aspera, rough blazing star, is an upright three- to five-foot-tall plant that will tolerate dry soil once established. It gets its name from the hairy texture of the leaves. It blooms in late summer and early autumn. Like all liatris, it is a great wildlife plant.

Liatris microcephala, smallhead blazing star, is a compact plant rising only 18 to 24 inches. It does best in well-drained soil and works well in rock gardens. This feathery-flowered purple liatris looks good at the front of a bed. It has a greater impact when planted in small clusters than when planted singly. Plant it in interspersed groups between other drought-tolerant plants, such as low-growing dianthus (*Dianthus*). It blooms in mid to late summer in zones 6 to 9. This is the least-hardy liatris.

Liatris pycnostachya, prairie blazing star, is tall, ranging in height from three to five feet. Plant it toward the back of a flower bed, especially if you have a fence for it to lean on. The midsummer blooms are usually purple-violet; there is also a white cultivar sold as 'Alba'. If you have clay-based soil, provide drainage by adding extra grit to the planting hole.

Liatris scariosa with a butterfly // Close up of *Liatris spicata* // White *Liatris spicata* with a bee // *Liatris spicata* 'Kobold' growing in my dear friend Pam Hubbard's pollinator garden.

Liatris scariosa, tall blazing star, can be three or more feet tall with long flower spikes that make a visually dramatic plant. It blooms in late summer and early fall.

Liatris spicata, spike blazing star, is most often used in gardens because it can grow in regular garden soil, and even in dryish soil once established. Water is needed during the growing season. Like all plants in this genus, it will die if it has wet soil in winter.

Various cultivars of this species are available. The straight species has more height variability than the cultivars. Look for the popular cultivars 'Floristan Weiss' and 'Floristan Violett' (also sold as 'Floristan White' and 'Floristan Violet'). With tall stems that rise up to three feet, they look lovely among surrounding perennials. 'Kobold' is a commonly grown cultivar that is shorter than the straight species. Growing two to two and a half feet tall, this sturdy plant does not need staking. It bulks up over time to make good-looking clumps. 'Kobold' is a great cut flower that is a nice vertical accent in a vase. It is a good choice for container plantings.

Lilium

lily

Full sun // Moist, well-drained soil // Zones 4–9 // Height 24–108 in. // Season of interest—late spring into fall

About This Bulb

Lilies are justifiably one of the most popular summer garden plants. With a wide variety of colors, shapes, forms, and heights, there is something for every garden. There are delicate slender lilies that look good at the edge of the woods, midsized lilies that look great in pots, and massive blooms that are bold beauties for the back of a flower border. Lilies look equally good in an informal cottage garden or a grand formal flower bed. If you love lilies in your garden, with careful selection of different species and cultivars, it is possible to have one or two in bloom from late spring until early fall.

The choice of lilies is bewildering. There are hundreds of useful species and hybrids to choose from for your garden. Before you buy a lily, make sure that you can match its sun and soil needs. Their flowering height varies widely, so pay close attention to this before planting. Then look at floral characteristics, such as color, flower shape, size, scent, and bloom time.

Lily flowers have six petals: three inner ones, and three outer ones, which may differ from the inner ones. Flowers can be trumpet- or bowl-shaped, flat open, or even have swept-back petals. The flowers can face upward, outward, or downward. The size of each flower can be petite, like some species lilies, or massive like 'Stargazer' or 'Scheherazade'. Some lilies have obvious spots, dots, or mottling on the petals. These markings are used to guide pollinators to the center of the flower. For gardeners, this adds to the decorative effect. The spots can be subtle or in a highly contrasting color. In a few lilies, the dots are raised lumps called papillae. If the lily is light in color, the spots are usually dark. Typical spot colors are maroon, brown, dark purple, black, or green.

Another ornamental lily feature is the presence of highly colored anthers. The pollen on the anthers can be bright red, orange, yellow, or brown, and is usually very sticky. If you have brought lilies inside to be used as a cut flower, you will know that the pollen can stain your fingers, clothes, or your best white tablecloth. Some flower arrangers cut off the anthers if staining might be a problem, for example in a bridal bouquet. There are lilies that are bred to have no stamens, to be used just for this purpose, including a light pink, spotted lily called 'Corsage'. These flowers are of no interest to insects, so they should not be used in a pollinator garden.

Not all lilies are scented, but those that are make their presence known in a garden from quite a distance. I have been to gardens where I am brought to a full stop by a strong waft of fragrance. It often turns out to be coming from a patch of lilies. The amount of fragrance depends on the time of day and the number of lilies in bloom. Different lilies have different scents, and those fragrances seem to be highly linked to color. White, light-colored, and bicolored blooms are usually highly fragrant; dark red seems unlikely to be really fragrant.

Planting and Maintenance

Choosing a suitable site to plant your lilies is one of the ways to guarantee success. Lilies are a classic example of a bulb that needs the soil to be moist, while at the same time well-drained. They also really do best when their bulbs are protected in shade, while their tops are up in the sun. Most lilies grow well in acidic to slightly neutral soil. There are a few like *Lilium henryi* that do best in an alkaline soil. The other thing to consider when choosing a planting spot is lilies can get eaten by herbivores, so it is good to give them some protection. Plant them in a fenced-in area, if you have one, or site them in a bed near your house.

Lily bulbs do not have any protective outer coating, so they dry out easily. They are best bought from a reputable dealer who will ship them to you freshly dug. Plant them as soon as you can. If you are storing them, keep them cool and lay some moistened paper over them to prevent them from drying out.

All lilies require well-drained soil. A method of providing this is to plant them in raised beds or in pots. The lily bulb hole should be at least three times the height of the bulb. Six to eight inches is about right for most bulbs. Stem-rooting lilies need to be deeper than ones that do not stem root. All lilies root from below the bulb, but some lilies also root from the stem that emerges from the top of the bulb. This trait means that they need to be planted extra deeply. The stem roots help absorb water and nutrients, and also anchor the growing plant. Good stem roots stabilize really tall lilies and protect them from rocking in the wind. *Lilium candidum* is planted shallowest at about an inch or two. Asiatic lilies are also not planted too deeply. Space lily bulbs about six inches apart to give them room to grow.

Amend the soil that will be around the lily by adding grit or coarse builders' sand to the bottom. Place the lily bulbs into the hole, pointed end up, sitting on the grit. Refill the hole with a mixture of two parts compost-enriched soil and one part grit. Firm the bulb in place, using your hands to remove air pockets from the soil. Water the lily to start the growth of its roots. Adding plenty of organic matter to the soil, and adding more before winter, will help the lilies survive the cold and improve the soil at the same time.

Plant bulbs in autumn, if they are available, or as early in spring as you can, so that they can get their roots out into the soil. Some species and cultivars are particular about when they are planted. If the lily is not hardy in your area, or if you are growing it in a pot, it is best to buy bulbs in spring. If they are bulbs that need to grow an extensive root system before they will flower, they are best planted in the autumn. In cold climates, if you know you will receive them after the ground is frozen, pre-dig the holes, protecting them with a waterproof covering, followed by some insulating organic matter such as leaf mold. Keep some potting mix with added grit in a frost-free area. Use this to fill the hole after planting.

Lilies have a fast growth rate, with one stem emerging from each bulb. Lily bulbs need to be well watered so that the stem and flowers can expand rapidly. At least 70 percent of each cell is water. If you feel that they require an additional boost, provide a high-potassium plant food throughout the growing season. Do not overfeed because that can lead to overly long stems that need staking.

Lily stems are usually strong enough to self-support, especially if you grow them in full sun. In windy gardens, grow lilies against a fence, or plant them to grow through lower-growing plants that will provide support. You can use stakes or plant supports, but they can detract from the overall display in the flower bed. Single supports with one ring attached to a metal stake can work if hidden by surrounding plants, as long as the lilies are not so tall that the stem snaps. It is helpful to put in a stake at the time of planting while you know where the bulb is located. A small stake can be used initially; if necessary, it can be removed later and replaced by a tall, sturdy stake.

Lilies usually bloom well the first year after planting, but some types get better after a few years. This especially seems to be the case with the tall lilies, sometimes referred to as tree lilies. The first year, they may be three feet tall, the next year four feet, and eventually taller. The lily bulb that you buy has to produce the whole stem and many flowers. As it settles down in your garden, it develops a bigger bulb, with more roots, and the ability to push out an eight-foot stem in one growing season.

Almost all lilies make great cut flowers. If you cut a lily and want it to flower again next year, only cut off as much of the stem and leaves as you need to make your arrangement. The rest can be left to grow and recharge the bulb.

Design Ideas

Lilies thrive when planted in a border full of other flowering plants which shade the lily bulb and roots from the strongest sun. Surrounding plants are usually shorter than tall lilies, so the lily gets the sunlight that it needs to grow sturdily. Lilies are a great addition to a mixed flower bed because they do not take up much horizontal space. The bulbs can be slotted in between neighboring perennial patches and will rise above them the next growing season. Lilies are lovely to plant next to a fence, where their long stems can lean over the structure. Grow them to come up through a nodding clematis, such as the reliable 'Betty Corning', for an echo of shape but a contrast in color.

When planning a garden design, add tall, bold lilies for drama. Look at the shape and colors of the lily and then find some floral companions that will complement them. Good accompanying plants for lilies are cool-season annuals like love-in-a-mist (*Nigella*), cornflowers (*Centaurea*), honeywort (*Cerinthe*), and then for later in the summer, warm-season annuals, such as cosmos (*Cosmos*) and zinnias (*Zinnia*).

If you would like to decorate your deck or patio, use a series of different lilies in pots. Choose by color, type, and whether they have a fragrance. They are easy to plant up ahead of time. Keep them at the back of your pot display, and when they are in bloom, rotate them to the front. Lily

flowers can be one color, but often have a second color toward the center. This can be visible on the outside, on the inside, or both. Lily flower colors are white, cream, light or dark yellow, peach, orange, red, burgundy, maroon, and light and dark pink. Sometimes the colors wash together to produce a watercolor effect. There may be a contrasting midrib vein that runs down the middle of the petal.

Cultivars

Asiatic hybrid lilies are early-blooming lilies that have been bred to produce easy-to-grow, sturdy plants. They work well in sunny garden beds or pots. Asiatic lilies are short, so they should be planted toward the front or middle of the flower bed. Their flowers tend to point upward, so that they catch your eye as you walk through the garden, making it easy to admire the color and markings. They are widely used as a cut flower.

Lily plants that are already in growth are often available for purchase in the spring. They can be easily added to a new flower bed or used to jazz up a lackluster one. When planting them, cluster three or five bulbs to produce a sweep of flowers. If the bed is large, repeat groupings of the same bulbs farther along. Try to resist the temptation to buy different Asiatic lilies and plant them in the same bed. It will certainly be diverse and colorful, but not harmonious. You could plant the different types, each in their own pot. When they are in bloom, you can move the pots around to make pleasing combinations.

A huge number of Asiatic lilies, are available. Look for ones that you like by color, height, and whether the flower is double or single. 'Corsage' is a lovely choice for pots and is prized as a cut flower because it has no pollen. For white-flowering Asiatic lilies, try 'Apollo' or 'Mont Blanc'. For yellow flowers, consider bright yellow 'El Divo' or 'Yellow

CLOCKWISE FROM TOP LEFT: *Lilium* Pink Perfection Group // *Lilium* African Queen Group // *Lilium* 'Black Beauty' // *Lilium* 'Conca d'Or' // Lilium 'Red Velvet'

Cocotte'; 'Istanbul', which is a lovely light yellow with slight markings with an outward-facing flower; and the even lighter, creamy 'Sweet Surrender'. For some bolder colors, 'Forever Susan' is a distinctive lily with dark burgundy-and-orange zones, 'Red Velvet' is red, 'Orange Summer' is bright orange, 'Heartstrings' has a yellow center and pinkish maroon petal tips, and 'Brushmarks' is orange and dark red.

Trumpet lilies live up to their name with dramatic trumpet-shaped blooms that stand out in the flower garden or in a container. They flower from midsummer onward. For a good pink, try Pink Perfection Group, which are scented and grow to five or six feet. 'Royal Gold' has strong yellow flowers. African Queen Group lilies have dramatic yellow-orange flowers with peachy apricot stripes down the center. They are four to six feet tall and bloom in summer.

Oriental trumpet lilies, Orienpets, are hybrids that are known for their large flowers, strong stems, and nice fragrance. They bloom in the summer. 'Anastasia' has fragrant, downward-pointing white flowers with a dark pink, star-shaped central pattern. It grows about three to five feet tall. 'Black Beauty' has beautiful dark red-pink flowers with a fine white edge and center. It flowers in midsummer. Reliable favorite 'Conca d'Or' has many flowers that are held on self-supporting four- or five-foot stems. The flower centers are yellow with white around the edges. 'Scheherazade' is a wonderful, extremely tall lily that produces many dark pink and greenish yellow flowers per stem. It blooms on stems that are eight feet or more and gets better every year. 'Silk Road' is a popular fragrant lily with dark maroon-pink centers and pink to white petal edges. 'Sweet Desire' is also fragrant; it has creamy yellow flowers with maroon speckles in the center.

CLOCKWISE FROM LEFT: *Lilium* 'Scheherazade' // *Lilium* 'Silk Road' // *Lilium* 'Sweet Desire' // *Lilium* 'Casa Blanca' // *Lilium* 'Muscadet'

Oriental lilies are some of the last to bloom in the season, flowering late in the summer into early autumn. The large, flat-faced flowers are usually outward facing. There are some double lilies in this group, such as light pink 'Lotus Queen'. All of these lilies perform best in acidic soil that contains plenty of leaf mold and composted organic material. 'Casa Blanca' is one of the best in this group with large, scented, pure white single flowers. The flower is bowl-shaped with petal tips that curve backward and orange anthers. It blooms in late summer and grows three to four feet tall. 'Garden Party' has a white flower with an obvious golden orange star at the center and some speckles. It is one of the shortest and a good one for growing in pots. 'Muscadet' is a lovely white to light pink

flower, freckled with darker pink, and orange anthers. There is also a short selection that is a good choice for container growing because it only reaches one and a half to two feet tall. 'Firebolt' is one of the darkest with upward-facing flowers that are scented. Delightfully fragrant 'Star Gazer' blooms in various shades of pink.

Species Lilies

Lilium candidum, Madonna lily, is a gorgeous lily that has been grown in gardens for centuries. It is perfect in any flower border and can take alkaline soil. It roots from the bottom, and not the stem, so it needs to be planted only an inch or so down in the soil.

CLOCKWISE FROM TOP LEFT: *Lilium canadense // Lilium formosanum // Lilium* 'Madame Butterfly' *// Lilium henryi* 'Lady Alice' *// Orange Lilium lancifolium // Lilium leichtlinii*

Lilium canadense is an elegant lily. It has small, recurved flowers that are often yellow on the outside and orange on the inside. Dark spots appear on the inside of the midsummer flowers. Growing three to eight feet tall, it can have up to 20 pendulous flowers on stems with whorled leaves.

Lilium formosanum, Formosa lily, is a long trumpet-shaped lily that resembles the earlier-flowering *L. regale*. Its base color is white, with a flush of pinky purple in a stripe on the outside. The stamens protrude slightly from the trumpet. The straight species can grow from four to seven feet tall. I grow a short selection that is self-supporting and very prolific with its seed production. It blooms in late summer at about

three feet. This lily grows easily from seed, either sprinkled directly into the garden or in pots. Scattering the seeds onto gravel is a great way to encourage germination. It will take a year or two to get to blooming size.

Lilium henryi is a charming lily that has very recurved orange flowers and long arched stamens. The petals have a bumpy surface due to the many papillae. This species lily is a good candidate to grow up through perennials and annuals. It blooms mid to late summer and can take partial shade, especially around its roots. 'Madame Butterfly' is a lovely cultivar. We have it growing up through a nodding purple clematis called 'Rooguchi'.

Lilium lancifolium, tiger lily, is one of the easiest to grow. This lightly fragrant tiger-colored lily is, of course, orange like its namesake animal. A characteristic of this fantastic lily is that it has lots of dark spots all over the petals. I love this three- to five-foot tall lily. I received my first ones as pass-along plants from a neighbor. They are easy to grow from the mature black bulbils that can be harvested from the leaf axils. When planted, each one takes a few years to grow to flowering size. Tiger lilies grow well among loose-growing herbs like feverfew and tansy (*Tanacetum*), and upright perennials like bear's breeches (*Acanthus*). As they grow, they are supported by later-blooming perennials, such as asters, that are only in leaf when the lily is in bloom.

Lilium leichtlinii has downward-facing flowers that are a lovely golden yellow with prominent dark spots. It grows to about four to six feet and blooms in late summer. Plant this lily in part shade and add plenty of compost to the planting area.

Lilium martagon, martagon lily, is the lily for people who do not even like lilies. Each three- to six-foot stem has many delightful little nodding flowers with petal tips that turn upward at the edges. The early summer flowers can be red, yellow, white, or orange, often with contrasting speckles and prominent orange anthers. It is regularly visited by hummingbirds and butterflies. Its slender stems hold whorls of leaves below the flowers. The narrow plant takes up very little space in the garden. It will grow in part shade with good woodsy soil that has plenty of leaf mold incorporated. It is hardy in zones 3 to 8. Martagons are lovely planted to come up through a groundcover. 'Claude Shride' is deep crimson-maroon with dark and white accents and yellow anthers. 'Guinea Gold' is another favorite of mine for its fabulous early golden yellow flowers with darker dots.

Lilium pumilum, coral lily, is a short lily that grows to about two feet. The bright orange-red flowers are small and dangling. It looks good when grown among perennials and annuals in a mixed flower bed.

Lilium regale, regal lily, is one of the first lilies to bloom in my garden. Before the flower opens, it looks as if it will be a light pinky purple. The flowers then open to reveal classic trumpet-shaped lilies with glistening white petals, a central pink stripe, and a yellow throat. Its bright yellow-orange stamens are held inside the flower and are colored by very sticky pollen. This lily is about six feet tall when established. The first year or two, it may not grow as tall, and some are naturally short. It is a reliable favorite for early summer gardens or in large pots. Combine regal lilies with roses, clematis, and summer perennials.

CLOCKWISE FROM LEFT: *Lilium martagon // Lilium pumilum // Lilium regale*

Lycoris

lycoris, surprise lily

Full sun to part sun // Moist, well-drained soil // Zones 6–10 //
Height 18–30 in. // Season of interest—summer into fall

About This Bulb

Lycoris are entertaining, slightly tender bulbs to grow for summer into fall bloom. They are unusual because the flowers appear in the summer, and then die down. You might wonder what has happened to their linear or strap-shaped leaves. These either grow after the flowers fade or wait until spring to emerge. The foliage then disappears, and the next summer the flowers pop back up. This strange double-growth pattern has given rise to the common names of surprise lily and magic lily. The flowers have six petals and prominent anthers that are a decorative feature. They thrive where summers are dry and warm. These bulbs are not bothered by herbivores. Lycoris can be used as cut flowers.

Planting and Maintenance

Plant lycoris bulbs as soon as you receive them. When planting them in the ground, cover the bulbs with six to eight inches of soil. They require good rich soil with excellent drainage. I have several species growing in my dry garden, where the top-dressing of rounded gravel acts as a summer mulch and insulates the bulbs over winter. Lycoris can be grown from seeds, but they resent being dug up and divided. If they are growing well, leave them in place.

If you grow *Lycoris* in a container, plant the bulbs at, or just under, the soil surface. Water them during the growing season and stop watering when the flowers are dying down and the bulbs are entering their winter dormant phase. Store the bulbs in their container in a place that is above freezing. Bring the bulbs out again next spring.

Design Ideas

Where lycoris bulbs survive in the ground, they can be planted to make sweeps of flowers in naturalistic groups, coming up through weak lawn, at the edge of shrub borders, growing up through groundcovers, in a rocky or gravel garden, or in a flower bed. Lycoris can be combined with surrounding perennials or annuals, as long as you give the bulbs enough space to get baked by the summer sun. Ensure that they do not sit in wet soil.

Lycoris make good potted plants. Grow them in moveable pots because they do not bloom all summer. When they have no flowers, they can be moved to a place out of sight, and then brought out to your display when they are in bloom.

Lycoris chinensis // Lycoris sprengeri // Lycoris radiata

Species and Cultivars

There are several lycoris species that are grown in gardens. Check their hardiness ratings before deciding whether to grow them in the ground or in a container. *Lycoris squamigera* is the hardiest bulb and is most often grown. The beautiful flowers look rather like a lily and its common names reflect this. Colors vary from a light shell-pink to darker pinks. The flowers are often fragrant. Other garden-worthy species include white *L. longituba* and creamy yellow *L. caldwellii*. They are hardy from zones 5 or 6 to 9, depending on how much winter protection they have.

Lycoris chinensis is hardy in my zone 6 to 7 garden. It has rich golden yellow flowers that emerge in late summer. They are an unusual color. Because of that, we have enjoyed combining them with great blue lobelia (*Lobelia siphilitica*) in both white and blue, milkweed (*Asclepias*), and Stokes' aster (*Stokesia*).

Lycoris sprengeri is another incredible-looking flower, hardy in zones 6 to 10. If you have never seen this flower in person, it is well worth seeking out. Even the best photos cannot do it justice. The colors are the true definition of sky blue–pink, and they change during the day as the light alters. It has long, dark stems.

Lycoris radiata, red spider lily, is less hardy than the others mentioned here. It grows well in the ground where winters are warm and in a pot in cold climates. The flowers are an unusual reddish color. The flower shape resembles a firework display, with petals that burst out from the center and curve back toward the tall stem. It is hardy in zones 7 to 10. Over time, it can form impressive patches in weak lawn.

Nerine

nerine

Full sun // Well-drained soil // Zones 8–10 // Height 18–24 in. // Season of interest—late summer into fall

About This Bulb

Nerines have lovely starburst-shaped flowers that come into their own in late summer or early autumn. The spicy-scented, funnel-shaped, pink or white flowers rise up on long leafless stems. They have slim, elongated petals that can be decorative for weeks in the garden. The strap-like leaves grow in spring and then fade away, sending energy to the bulb to power their flowering later in the growing season. They need a long growing season to flower well. Nerines are pest resistant.

Planting and Maintenance

Plant nerine bulbs in a sunny site with well-drained soil. Good places to plant them include raised beds or a position in front of a sunny wall, where heat is radiated back onto the bulbs. They perform best where the bulbs get a good summer baking. Plant bulbs about four inches deep in spring, after the soil has warmed up. In cool areas, where nerines are borderline hardy, plant the bulbs deeply for extra winter protection, and add a thick mulch of fluffy material like pine needles or salt hay. Plant the bulbs about six inches apart. Nerines will need plentiful water during the growing season and may need feeding with high-potassium fertilizer.

In gardens where the bulbs are not fully hardy, grow them in a pot. Plant with the neck of the elongated bulb exposed and an inch or two apart for good show. Even when grown in pots, nerines need a hot position in full sun in the summer. The pots will need to be brought inside for the winter. This is when you reduce their watering, to let them have a period of dormancy before starting to water them again in the spring.

Nerines do well when their roots are constrained. However, they will stop flowering if they are truly overcrowded. The bulbs do best when undisturbed. They will bulk up to form a clump that may push themselves out of the ground after a few years. This is when you know that they need dividing. Other signs are when they have filled the pot with roots, or you see the bulbs fail to flower. Take the whole mass of bulbs out of its container while they are dormant, divide up the bulbs, and repot. Get rid of any bulbs that look soggy, diseased, or damaged. Small bulbs take a couple of years to reach flowering size; large bulbs will flower the next growing season.

Nerine bowdenii // Nerine bowdenii

Design Ideas

Nerines are perfect bulbs to grow in a warm climate, where they make great late-blooming flowers in a sunny, dry, protected spot. Combine them with other late-season flowers, such as salvias (*Salvia*) and penstemons (*Penstemon*), or plant them among low-growing plants in a rock garden. Carefully place surrounding plants away from the nerine bulbs to ensure they are in full sun. Nerines make good cut flowers.

In cooler areas, nerines are wonderful container plants. The flowers are slender and wispy, so plant several of the same cultivar in a pot for best impact. For a late summer vignette, combine pots of nerines with containers of agapanthus, amarines, tuberoses, and dahlias.

Species and Cultivars

Nerine bowdenii is the best nerine for most gardens; in colder areas, it will do well in containers. The flower colors of nerines are dark pink, light pink, or white. Cultivars vary by flower color and have slightly different heights. *N. sarniensis*, Guernsey lily, has been grown in greenhouses for many years in temperate or cold areas. *N. undulata* has lovely bright pink flowers. The hybrid ×*Amarine* is a cross between *Nerine* and *Amaryllis belladonna*.

Polianthes

tuberose, polianthes

Full sun // Moist, well-drained soil // Zones 8–10 // Height 18–36 in. // Season of interest—late summer into fall

About This Bulb

Polianthes tuberosa, tuberose, is one of the most delightfully fragrant summer bulbs that you can grow. Once you have smelled a tuberose, you will know why everyone needs to grow them. The fragrance is haunting, memorable, and hard to describe. They are used as a cut flower and in the perfume industry. The flower is waxy in texture, white, pale pink, or occasionally a yellow-cream color. The flower shape is single or double. The tubular flowers are arranged pleasantly at the top part of the long single stem. The leaves are grassy and silvery green. You may also see this plant listed by the Latin name *Agave amica*.

Planting and Maintenance

Plant tuberose bulbs about four inches deep in spring, after the soil has warmed up. They need a sunny area of your garden to get the necessary summer baking to flower well. Consider adding some grit or gravel on top of the soil above where you planted the tuberoses. This will help heat up the soil below. They also need a long growing season, so you could start their growth inside and then plant the bulbs out once they have sprouted. They require a period of winter dormancy. After the leaves begin to die down, dig the tuberoses up and store them inside in a frost-free place until next spring. In climates with cool summers, it may be best to grow tuberoses indoors as a potted plant or find a hot microclimate outside.

Design Ideas

Choose a full sun site for your tuberoses, that is near where you will walk regularly, to enjoy the fragrance. Tuberoses are tall plants that are perfect for the middle of the flower bed, or for a large container. Make sure not to overcrowd them, so the bulbs can still get plenty of sun. They can be planted in a cutting garden. They are great to add to an arrangement because of their scent.

Species and Cultivars

Polianthes tuberosa, tuberose, is often a single white flower but is available in several other cultivars that vary in color and shape, while still retaining the all-important fragrance. One of the most widely grown double tuberoses is called 'The Pearl'. It has another layer of petals within the normal tubular flower. Pink cultivars to look for include single-flowered 'Sensation' and the double 'Pink Sapphire'. For a light yellow, choose 'Super Gold'.

Polianthes tuberosa

Sternbergia

sternbergia, fall daffodil, autumn daffodil

Full sun // Well-drained soil // Zones 6–9 // Height 6 in. //
Season of interest—late summer into fall

About This Bulb

Sternbergia are the final burst of garden sunshine that brings the outdoor bulbous year to a close in my garden. Its egg-yolk-yellow, chalice-shaped bloom is dazzlingly bright on a sparkly autumn day. They are a nice surprise in autumn, as these flowers look like they should be blooming in spring. The flowers emerge first, followed by bright green, strap-shaped leaves that persist throughout winter.

Planting and Maintenance

Plant sternbergia bulbs about four to six inches deep and four inches apart in a sunny, well-drained spot that receives a summer baking. To do well, they need their own area that is not overshadowed by summer-growing perennials. To help increase heat absorption into the bed, top it with gravel. In climates where summer sun is unreliable, look for an area of your garden that is a sheltered suntrap.

Design Ideas

The ideal site for sternbergias is in a sunny, raised rock or gravel garden. They look great when planted around a nice bench that catches the warm autumn sun. Combine them in an area that also has spring bulbs, like yellow crocuses and winter aconites, for a continuation of small bulbous plants from spring through fall.

Species and Cultivars

Sternbergia lutea is the species that is grown in gardens. It is known for its fall-blooming, sunny yellow, crocus-like flowers and green leaves.

Sternbergia lutea

Tulbaghia

tulbaghia, society garlic

Full sun // Well-drained soil // Zones 7–10 //
Height 12–18 in. // Season of interest—summer

About This Bulb

Tulbaghia plants are used in gardens as summer-blooming plants that have fine, narrow leaves. The blooms are clusters of purple flowers borne at the top of narrow stems. Tulbaghia is a frequent addition to sunny terraces and herb gardens.

Planting and Maintenance

Tulbaghia, society garlic, grows from bulb-like tuberous roots, but is often purchased as a potted plant. In many gardens, it does best in containers, due to its need for very well-drained soil. Include extra grit in the planting mix when repotting. Tulbaghia needs little maintenance if situated in a warm, sunny place. Where it is not hardy, bring it inside for the winter.

Design Ideas

In warm winter areas, tulbaghia makes a great dry garden plant. It can be used to create an edging along paths, in gravel gardens, or as a bedding plant. In colder areas, tulbaghia is normally grown singly, in its own container, which can be brought into a frost-free place for the winter. In the ground or in pots, combine with other plants that like sunny dry places. Tulbaghia fits in nicely among silver-leaved herbs like lavender (*Lavandula*), sage (*Salvia*), and santolina (*Santolina*), where this grouping makes a vignette of silver, light green, and purple.

Species and Cultivars

Tulbaghia violacea has green leaves and light lilac flowers. Some cultivars have variegated leaves that add a little more interest to the plant.

Tulbaghia violacea

Zantedeschia

zantedeschia, calla lily

Full sun to part shade // Damp to wet soil // Zones 7–10, varies // Height 24–36 in. // Season of interest—summer

About This Bulb

Calla lilies are grown for their distinctive-looking, long-lasting flowers. What we think of as one flower is actually a collection of flowers—this makes up the central protuberance called a spadix. The surrounding part that looks like a petal is the spathe. The spathe can be in a variety of different colors, but the classic color is white. The arrowhead-shaped leaves are also decorative, and some have mottled markings. Different types are either evergreen or deciduous.

Hardy calla lilies are perfect in the ground in wet soil, but some of them can take average-to-moist soil. They are grown for their leaves and their summer bloom. The tender ones are often used as container plants for their range of flower colors, such as cream, mango, maroon, pink, and yellow. These bulbs are resistant to herbivores. Calla lilies are also grown as cut flowers and are a favorite for wedding bouquets.

Planting and Maintenance

If the calla lilies that you are growing are hardy for you, plant them directly into damp to moist soil, at the depth of about two or three inches. Some species can grow in standing water or boggy areas, and others are fine in good garden soil.

When growing calla lilies in containers, plant the bulbs two to three inches below the surface. The concave side is the top, but if you cannot tell, put them on their sides. Start them inside, in late winter, in a frost-free place. Water them sparingly while you are first getting them to sprout. After the last frost in spring, gradually acclimatize them to outdoor temperatures.

In the autumn, they can be brought inside to be stored in their pots over winter. At this point in their lifecycle, reduce the amount of water that you give them. The leaves will then die back and fall off. Keep them in their dormant condition, slightly moist but not wet, in their pots, in a frost-free area for the winter. Inspect the bulbs in early spring. They may have started growing. They can be taken out of the pot and repotted in fresh potting soil. Calla lilies may be eaten by Japanese beetles.

Design Ideas

Hardy calla lilies are used in boggy areas, such as stream banks and the margins of ponds. They can be used in the moist soil at the bottom of a

Zantedeschia albomaculata // Zantedeschia 'Sun Club'

downspout or in a rain garden. Grow them with water-loving irises. The water tolerance varies by species, with some being able to grow in regular soil in mixed flower beds, while others can be planted into a pond planting container and sunk into water. I have some that are hardy in my garden, that grow in regular soil with moderate moisture. The tender species are perfect for growing in containers. They usually look best planted with one color to a pot because the colors are quite strong. Their distinctive shape is a great addition to a summer bulb vignette.

Species and Cultivars

Zantedeschia aethiopica is grown for its unspotted leaves and large white flowers. It is able to withstand boggy soils. It reaches two to three feet tall. *Z. albomaculata* are calla lilies with white flowers and lightly spotted leaves.

Tender calla lilies are hardy in frost-free climates. In most gardens, they are grown in containers. They are available in a wide range of colors, so look for cultivars that suit your design theme. 'Sun Club' has bright yellow flowers and leaves with little white dots; 'Odessa' has really dark purple-maroon flowers and mottled leaves; 'White Flirt' is a creamy white; 'Morning Sun' is orange; and my favorite, 'Zazu', has pink flowers and mottled leaves.

Zephyranthes

rain lily, fairy lily, zephyr lily

Full sun // Regular to moist soil // Zones 7–10 // Height 8–12 in. // Season of interest—summer into fall

About This Bulb

Zephyranthes, or rain lily, is a bulb that flowers after rains, usually in late summer or early autumn. It can reflower after a period of dry weather, followed by some good rains. Crocus-shaped flowers borne on flexible stems rise above the bright green, grass-like foliage. Rain lilies are suitable for growing in the ground or in pots. They are American bulbs that comes from the south and southwest of the country. They are resistant to pests.

Planting and Maintenance

Plant *Zephyranthes* bulbs about three inches deep, either in the ground or in a pot. If they are not hardy, dig up the bulbs in autumn and over-winter them in a pot (in a similar way to gladioli). They can be dug and divided when the clumps get really congested, but they seem to flower best when packed together, especially in a pot.

Design Ideas

Rain lilies are wonderful at the front of a bed, so their beautiful flowers can be admired. Plant them in clumps or elongated groupings. Where they are hardy in the ground, they can spread happily. Rain lilies look great when grouped in a container, either alone or as part of a mixed design. Make sure the other flowers like the same sunny conditions. Suggestions for other plants to combine with these bulbs include nerines and colchicums. Rain lily flower colors are white, pink, and yellow.

Species and Cultivars

Zephyranthes candida is one of the most commonly grown rain lilies. It is also the hardiest. It blooms from summer into fall with pure white crocus-like flowers and yellow stamens. *Z. citrina* has yellow flowers and similar hardiness. *Z. atamasca* is a delightful white lily-like flower that is a common late spring flower in the southern parts of North America. *Z. grandiflora* is pink and only hardy in zones 9 to 10.

Zephyranthes candida // Habranthus robustus

Habranthus robustus is a closely related plant that looks like *Zephyranthes* and is also sold as a rain lily. It looks similar in shape and is used the same way in the garden or in a pot. It has pink flowers.

Once you have decided which bulbs you would like to grow, it is time to think about how you will combine them in your garden. A bulb flowering alone is delightful. However, integrating that same bulb with other bulbs, and the rest of your garden plants, brings out the best in all of them. Creating combinations that bloom at the same time and look good together is one aspect of bulb growing. The other part of the selection process is establishing wave after wave of bulbs that flower in sequence. The resulting displays will bring you delight during the course of the growing season.

A densely planted bed of many types and colors of tulips at the Keukenhof gardens, near Amsterdam, Netherlands.

PERSONALIZING YOUR BULB GARDEN

When faced with all of the lovely possibilities, it can be difficult to decide which bulbs to plant. Focusing on why you want to grow bulbs in your garden can help you narrow down your possible choices. Maybe adding beauty to your garden is your primary motivation. Perhaps you love fragrance and want to add scented bulbs. Other gardeners may be inspired to extend their flowering seasons. Some people are drawn to certain colors, and bulbs can contribute amazing hues to the garden picture. Whatever your reasons for growing bulbs, it is useful to think about how you will use them in your garden and ways to arrange them for best effect.

Finding Places to Plant Your Bulbs

When you are planning your bulb plantings, begin with a walk around your garden. Bulb flowers provide a wonderful sign of the changing seasons, so they should be planted where you spend time in the garden or can see them from a window. You can use cheerful bulbs to greet you right by your door, gate, path, or mailbox, where they will be charming to look at as you come and go.

Another good option is to plant bulbs close to places where you sit in your garden, whether on a patio, terrace, deck, or by a bench. Adding bulbs to a current flower bed is usually an easy choice because the soil is already prepared. You may be able to find little nooks and crannies throughout your garden where you can tuck in a few bulbs. Spaces around existing woody plants are another possibility. These small bulb pockets can be just as fun to create as big displays and are much easier to plant.

Planting Styles

Once you have found an area where you might plant bulbs, it is time to think about the style of planting you like. Whether you are planning a new garden or refreshing an existing one, bulbs can play an important role in adding color and interest. The way that bulbs are arranged in a bed produces very different looks depending on whether they are planted in a formal or informal style. In formal gardens, placement is symmetrical. The bulbs may be removed after they flower to be replaced by the next batch. Informal gardens use bulbs in naturalistic or irregular patterns. Home gardens are often semi-formal in style, with a mix of formal and informal areas.

Formal Style

Large formal gardens in some public parks may have patterned flower beds that contain closely packed bulbs. This theme often appears in historic gardens, especially Victorian ones. The flowers are chosen primarily for their color and form and are replaced several times during the year. The plants, when they are used in this way, are referred to as bedding plants. The beds are often geometric and may be surrounded by straight paths, clipped low-growing hedges, or stone edging.

While these extensive plantings are beyond the reach of most homeowners, they make wonderful places to visit. In your own garden, you can use the same principles by making small areas of formally bedded bulbs. Change out these bulbs several times a year. When closely planted bulbs are all in bloom, they make your garden bright and showy. Each seasonal changeover brings an opportunity to choose a new plant palette and color scheme.

To use bulbs as bedding plants in your garden, choose a flower bed and plant it densely with seasonally appropriate bulbs. In spring, tulips and imperial fritillaries are perfect to fill these beds. They can be underplanted with smaller bulbs like muscaris, or herbaceous plants like wallflowers or forget-me-nots. After they bloom, remove all the plants to make way for summer bulbs. These could include cannas, caladiums, bedding dahlias, or begonias, underplanted with warm-season annuals, such as zinnias (*Zinnia*), marigolds (*Tagetes*), or salvias (*Salvia*).

Planting Pockets

Tender bulbs, or ones that are treated as annuals in your garden, can be included in the flower bed in reserved areas called planting pockets. These designated areas in your bed are used for different seasonal bulb types during the year. In this way, you will always have bulbs in peak bloom. This is a more intensive method than just growing bulbs that are left in the ground year-round, but it is less work than formal bedding.

1. Place bulbs, such as this 'Red Highland' Asiatic lily, where you will be delighted by them as you walk along a path, like here at Chanticleer gardens in Pennsylvania. **2.** The way you situate bulbs in a bed can contribute to the sense of formality or informality. Here, *Allium siculum* and *A.* 'Purple Sensation' mingle in an informal bed. **3.** Tulips are a great candidate for bedding out to make blocks of color, as demonstrated here at Keukenhof, Amsterdam, Netherlands. **4.** A single red dahlia has been placed in a planting pocket in the middle of this perennial bed.

1. Informal beds look appropriate beneath deciduous trees, like this combination of erythroniums and tulips. **2.** Summer-blooming bulbs like 'Nova Lux' gladioli fit easily into mixed plantings with annuals. **3.** Incorporating bulbs into a mixed flower bed brings in additional shapes and colors. In the Long Border at Great Dixter in Sussex, England, byzantine gladioli and an assortment of alliums contrast with surrounding plants.

There are two ways to add bulbs to planting pockets. One method is to plant them directly in the ground. Another choice is to grow bulbs in pots that you sink into those spaces. A good sequential pairing is hybrid tulips followed by dahlias. Plant the tulip bulbs into the bed in autumn. After their flowering time is finished, remove them and plant dahlias in their place for summer through autumn bloom. Other options include daffodils for spring and lilies for summer.

Informal Styles

In contrast to formal bulb plantings, informal gardens use hardy bulbs that stay in the ground for years. These bulbs are arranged in irregular groups, with clusters varying in shape and size. They produce an unstructured look that resembles something that could be found in nature.

Bulbs can be planted informally in mixed flower beds, in shade gardens, among tall perennials in meadows, in prairie-style plantings, or in a lawn. Choose bulbs for these uses that are long-lived and good for naturalizing.

Bulbs in Mixed Flower Beds

A popular way to use bulbs in the garden is to add them to a mixed flower bed. These beds are "mixed" because they contain different plant types, including long-lived perennials, yearly annuals, bulbous plants, and (occasionally) shrubs. The flowers in these beds are integrated together to produce a lush, seasonal look. Bulbs bring excitement to the display with their parade of unique shapes and wide range of colors.

When making a new flower bed, you can include bulbs from the start. If you have an existing flower bed, you can add bulbs among the other plants. Either method requires some thought as to where the bulbs will be placed, how many bulbs will be planted in each group, and the quantities needed to make a good impact. When assessing which bulbs you want to add, think about what time of year you want flowers. There are different bulbs that you can use to have bloom from early spring through late fall.

Perennial plants that live for several years form the backbone of a mixed planting and are usually planted first. This group includes long-lived bulbs and herbaceous plants that are hardy in your zone. Tender bulbs for summer and fall are slotted in between the permanent plantings. These bulbs are temporary occupants that are used like annuals in the flower bed. Do not agonize about the choice or placement of the non-hardy plants because if you do not like them, you can change them next year.

Bulbs can be easily added to mixed beds. When you plant them, they take up very little horizontal space. As the bulbs grow, their stems push up through surrounding plants. The flowers of bulbs are a useful addition to the overall scheme. They have striking appearances that are different from most other herbaceous plants. Ideally, your mixed bed will be situated in full sun or afternoon shade, and it will have soil that is well-drained but rich in organic matter.

Arrangement of Bulbs Within the Flower Bed

Bulbs come in an incredible range of heights: little ones for the front of a bed, medium-height bulbs for the middle of the flower bed, and tall ones for the back. In a bed that is viewed from several sides, plant tall bulbs in the center of the bed.

When deciding where to plant your bulbs in the bed, look at the height of each plant at maturity. Short bulbs that are under a foot tall—such as scillas, muscaris, ipheions, and crocuses—are just right for the front of a spring bed. They are tiny treasures with miniature flowers that need to be seen up close to be appreciated. For the front of the bed later in the year, use short begonias, caladiums, and bedding dahlias that bloom at about a foot tall.

The central band of the flower bed is a prime location for lots of flowering bulbs. Visually divide the center of the bed into mid-front and mid-back. Plants that are just taller than one foot are perfect for the mid-front. Those that are two and a half to three feet tall suit a position in the mid-back. Anything taller than three feet is usually planted near the back. The exception to this advice is for slender plants like gladioli or liatris, which can be planted in any height band. You can see around them wherever they are planted.

In spring, short daffodils and tulips will be in the mid-front with taller types behind them in the mid-back so that they are all visible. For summer into fall, choose crocosmias, eucomis, caladiums, short lilies, and bedding dahlias, and place them according to their heights.

Flowering bulbs for the back of the bed do not gain their true height until late spring. Some of the tall fritillaries and alliums are the first to tower over the flower bed. From summer into fall, there are lofty plants such as lilies, cannas, dahlias, and tall alocasias. These height bands are not distinct stripes down a bed, they are general guidelines as you plan.

Bulb arrangement continues throughout the year. Plan to have a succession of different bulbs in the same bed, that can come up and bloom one after the other. Deeply planted spring bulbs can be overplanted with shallow-rooted summer bulbs or annuals that will use the same patch of ground later.

Repetition

Mixed flower beds and borders are a beautiful way to grow a wide array of plants. The prettiest flower beds look as if all the plants belong together, with no jarring elements. These beds are unified by the repetition of certain key plants, colors, or shapes.

When bulbs are in bloom, they have noticeable flowers. The repetition of recognizable bulbs in each season visually links your plantings together. To achieve this look, buy only a few types of bulbs, rather than a wide variety, but buy more of them. Use these bulbs in several places throughout your bed. Another way of achieving a harmonious look is to choose a color scheme for the bed.

If your mixed beds are semi-formal in style, the repeated bulb groups may form a distinct pattern. For example, if you have beds that flank a path, the plantings on the two sides could be mirrored. In a linear bed, you could include a series of bold plants that are set out at regular intervals. Informal planting schemes also benefit from repetition. In these beds, the clusters are not evenly spaced, and the number of bulbs per hole varies.

1. In this mixed summer flower bed, short agapanthus are planted toward the front, with layers of intermediate crocosmias and tall agapanthus interspersed behind. 2. A flower border in front of a wall in the Cotswolds, England, is planted with bright red tulips and muscaris coming up through bearded iris foliage. 3. The same border just over a month later is now dominated by this lovely blue-purple bearded iris with forget-me-nots underneath. 4. Repeating colors throughout a flower bed is a great way to coordinate the look, as seen here in this late spring planting at Chanticleer in Wayne, Pennsylvania.

Numbers of Bulbs Per Planting Hole

When making your plan for adding bulbs to a mixed flower bed, it helps to know how many you need to plant together in one hole. Large bulbs, such as cannas, lilies, alocasias, colocasias, tall fritillaries, and dahlias, are planted singly. Medium-height plants, like tulips and daffodils, look good when several of each type are grouped together. These are usually planted three or five bulbs per hole.

Small bulbs should be planted in groups of 10 or more in one hole. This is the easiest way to plant them, and they generate more of an effect in the flower bed when clustered.

Bulb Spacing Within and Between Groups

One of the hardest things to learn is how far apart to space bulbs. They usually come with suggested planting distances. Sometimes I follow this, and other times I do not. If you are growing bulbs for show, like large hybrid tulips which will be discarded after bloom, plant them much closer than the instructions suggest. This creates an abundant display. Imagine what your flowers will look like when they come up. Your reaction when they are in bloom should be "Wow, that looks great," not "Wow, I wish I had bought more bulbs."

To get maximum show with bulbs, concentrate your bulbs into a restricted planting area. If you spread out a hundred bulbs around your garden, the resulting show may not be as spectacular as you might like. However, taking the same hundred bulbs and clustering them into one area produces a great display when they are all in bloom. This advice applies to in-ground planting as well as when planting up containers. If you love the look

TOP: This cactus-shaped white dahlia is repeated in groupings along the length of this mixed flower bed.

BOTTOM: To create a full display, cluster small bulbs like these crocuses together in one hole as you plant them.

OPPOSITE: Ranunculus bulbs look great when gathered together in a grouping, but they need space to grow. They should each be planted in their own hole.

when they come up, repeat this planting at the same density in other beds in succeeding seasons.

When you want bulbs to naturalize, space them farther apart than is proposed so that you can leave them in the ground for some years before they need to be divided. Each bulb can get plenty of light, water, and nutrients. The bulbs will multiply, and the clump will need to be divided eventually. To increase the naturalistic effect, vary the number of bulbs per hole and the distances between each grouping. Creating a natural-looking appearance to your bulb planting is a difficult process because we gardeners are prone to planting in straight lines, with even spacing between plants. The old way of randomly arranging bulbs was to throw the bulbs backward over your shoulder and plant them where they landed. To me, this does not produce the effect that I want. I like to think of the bulb grouping as being akin to arranging a solar system, with planets, satellites, and moons. Each celestial type has a different-sized bulb grouping and is spaced apart from each other in orbits. The resulting bulb display is like a starry night, where the flowers have a relationship to each other that looks loose but thoughtful.

In mixed flower beds, the spacings will vary, depending on the size of the bulbs and the look that you want. For large bulbs plant a few in a hole; usually odd numbers look best. Uneven numbers of bulbs in a hole produces a pleasing, natural effect.

Think of an uneven triangle shape for three big bulbs, with one bulb at each corner. When planting three bulbs together, do not automatically plant them in the shape of an equilateral triangle. Try spreading the bulbs into an irregular isosceles triangle instead. When planted this way, each bulb grows up in close proximity

These dahlias are packed together closely within their row, but they have enough space between rows to grow well.

to its neighbors, but the resulting flowers do not look too bunched together. Four bulbs per hole never looks very good. No matter how you arrange them, the display looks square or boxy. If you buy a bag of four bulbs, plant three in one hole and the last one in its own hole close by. Five bulbs usually look good together and can be planted like the dots on a die. When the bulbs come up they should be visually related to each other, while at the same time having enough room to grow.

For tiny bulbs, I make clusters and clumps, where bulbs, within a grouping are close together. There is an obvious space between this clump and its neighbor. Dig one hole and put 10 to 20 little bulbs in the bottom of the hole, spaced out slightly. Each group is separated from its neighbor and repeated down the bed. The planting hole can be any shape you like. An elongated trench or a wave-like shape creates a ribbon of flowering

My "Side of House Bed" is a picture in midsummer with various lilies rising through perennials, including additional bulbs, such as short gladioli and crocosmias.

NORTHVIEW MIXED BULB BED

I have a bed along the side of my house which used to get short shrift when it came time to plant bulbs. In the past couple of years, we decided to add lilies for the summer. This experiment began badly with my purchase of a too-good-to-be-true grab bag of lilies—it promised one thing and produced another. The first lesson I learned was to buy from a reputable bulb dealer. Things started looking up with the addition of some old-fashioned tiger lilies from my neighbor and some new well-chosen lily bulbs.

The gardening year in this bed begins with snowdrops and a few crocuses. Next come daffodils that poke their way up through the catmint (*Nepeta*) and feverfew (*Tanacetum*), which provide a pest-resistant scented buffer between the bed and the lawn. Bearded irises have thrived in the sunny dry conditions in this bed. They bloom alongside the peonies (*Paeonia*) in late spring. The color scheme continues to be rather crazy, since I decided that I really needed scarlet red crocosmias in that bed. They look good coming up with bronze fennel (*Foeniculum*) and yellow rudbeckias (*Rudbeckia*).

By summer, the lilies begin to bloom and carry this bed for the next few months. They bloom in succession from the early non-fragrant Asiatic lilies to the excellent 8- to 10-foot stems of 'Scheherazade'. Adding more bulbs has elevated it from a bed that was rarely enjoyed, to one that is now worth visiting regularly.

bulbs, which looks like it is woven among the other plants. The repeated bulbs become a lovely unifying feature. Tiny spring bulbs can also be planted as mixed bulb groups with a selection of little bulbs together in one hole. Use a variety of colors and forms to find bulbs that will look good together when in flower. Choose from crocuses, muscaris, ipheions, miniature daffodils, and blanda anemones.

To work out how many bulbs I need to order, I walk alongside the bed and count how many clumps I can fit in. I take that number and multiply by the number of bulbs per clump. Usually, I am much less scientific and wing it. I might decide I need more muscaris or caladiums for a particular bed and order a bag or two. The good news is if you like the look of a certain bulb when they bloom, you can add more bulbs to the planting next year.

In the first year in your garden, bulbs may flower a little later than expected, as they are getting settled into their new patch of soil and growing roots. Perennial bulbs will flower earlier in subsequent years. You can use that knowledge to your advantage by planting a few more of your favorite type of bulb each year to extend the show.

Play off the colors of your house when choosing bulbs to plant nearby. These bronzy tulips reflect the color of the door and the new foliage of the roses behind them.

Bulbs for a Color-Themed Garden

Color in gardens elicits a strong personal response. Our like or dislike of certain hues or combinations is often one of the first reasons that we are attracted to or repelled by certain gardens. Bulb flowers come in an incredibly wide range of colors with something for every taste. You have your own preferences. If you incorporate them into your garden, they will bring you joy.

Think about the background color of your house, the fence behind the bulbs, or any other predominant colors in the vicinity. Looking at all the neighboring objects helps you plan a bulb planting that is in sympathy with its surroundings.

Colors for Spring Bulbs

The early spring color palette consists primarily of whites, purples, and yellows from snowdrops, crocuses, and winter aconites. By mid-spring, the colors get brighter and more diverse. The plethora of tulip colors alone can allow you to create combinations that are wild, subtle, or somewhere in between. You can add in other bulbs that flower in true blue like muscari, strong pinks or deep purples from hyacinths, orange-cupped daffodils, or rich red coronaria anemones or ranunculus.

Alongside these brilliant colors, there are plenty of subtly hued bulbs, like some fritillaries, that add incredible depth to the planting combinations. Those who like the idea of a white flower garden are in luck, beginning the year with snowdrops and white crocuses and continuing with beautiful all-white daffodils, leucojums, hyacinths, and tulips. The tail end of spring into summer is a colorist's dream time in the garden, with various irises, alliums, and late tulips flowering in a multitude of hues.

TOP: Choose your most impressive bulb first, then work out what colors you want to go with it. Here, the color of a yellow imperial fritillary is echoed in daffodils and tulips of a similar color, and is contrasted by red tulips and blue Juno irises.

BOTTOM: For this color scheme, the centerpiece is a copper-colored imperial fritillary with dark foliage. This is surrounded by white-and-orange daffodils and tulips in a range of complementary colors.

Colors for Summer and Fall Bulbs

In the summer and fall, bulbs are unlikely to be the only thing in bloom. Have fun choosing colors to be part of the lush, ever-growing garden picture. Bulb flowers come in a wide range of colors that can be added to your preferred color scheme.

At this time of year, include bulbs that are grown for their foliage, like cannas, caladiums, and colocasias that add bold splashes of color. The hot colors of tropical and subtropical plants like gingers and crocosmias add punches of orange, red, and yellow to a flower bed. If you prefer cool colors, look for liatris, lilies, gladioli, and dahlias in hues of pink, purple, and white.

ABOVE: Dahlias are prized for their wide range of colors, as seen in the Dahlia Dell, a showplace garden in Golden Gate Park, San Francisco, California.

OPPOSITE: Red lycoris and purple colchicum liven up the fall display garden at Brent and Becky's Bulbs, Gloucester, Virginia.

Planning Your Color Combinations

When assembling a bulb list for a flower bed, think about the effect that you are trying to achieve. Some people prefer a set color scheme, while others are happy to have flowers of any color in the same bed.

If a certain flower color is important to you, start with bulbs in that color. Work off this first pick, adding other bulbs to reinforce or complement your chosen color scheme. When you are beginning to grow bulbs, or if you are unsure about how to combine bulb colors, it is often best to stick to a color palette of two or three possible colors. As you choose your combinations, be aware that plant colors that surround your bulbs affect your perception of the actual colors. There is lots of scope for individuality.

The most soothing bulb combinations lean into blues or purples. These are cool colors that benefit from even a small amount of white or light blue to make the other hues stand out. In the spring, there are plenty of blue and white bulbs such as scillas, ipheions, and anemones. By late spring, the bulb colors shift toward purples with alliums being a highlight. Purple is also found in gladioli and dahlias.

If I were creating some bulb combinations, I would start with my favorite colors of pink and purple. I would probably choose a gorgeous bright pink lily as my first pick. Then I might add some gladioli in shades of pinky purple and white galtonia with white caladium on the shady end of the bed for summer. To bolster the show into fall, I would add some late-blooming Formosa lilies, acidanthera gladioli, and colchicums. In autumn, in the same bed, I might plant some bulbs of pure white daffodils and lovely pink hyacinths that would bloom the following spring.

Someone else who loves orange and white might begin the year with crocuses, ranunculus, tulips, and imperial fritillaries. They would then progress into summer with white irises and orange Asiatic lilies, followed by striped cannas, crocosmias, tuberoses, and dahlias in luscious sunset colors. If you wanted to change this color scheme, you could add in blues or purples to the mix to make the orange stand out. Another approach could be to reduce the amount of white by adding scarlet red or vibrant yellows.

A pure white garden is something that many people aspire to create. From personal experience, I can say that it is more difficult to do than it might appear. I have had success with white in spring with some delightful all-white daffodils, white 'Jeanne d'Arc' crocuses, and white or light blue muscari. Later in the year in that same bed, white alliums and camassias could be followed by pure white 'Casa Blanca' lilies that are amazing for both their looks and their scent. Any white gladioli are great, including the fall-blooming one that is called acidanthera. There are plenty of white dahlias, but their super green foliage tends to reduce the overall whiteness of the display. Developing a color scheme that is white, plus another color or two, is easier to create than a pure white garden. This gives you more choices and may end up pleasing you more. Green as one of your other color choices would bring a sense of freshness to the garden.

Yellow is a color that is widely found in bulbs from early spring through later autumn. The possible yellows include bright, hot yellows, yellows with a touch of green, and pastel yellows. Each one of these has a useful place in bulb color

1. This midsummer planting centers on a hot-colored crocosmia, surrounded by annual marigolds. **2.** The rich purple of this allium stands out against the bright ladybird poppy (*Papaver commutatum*). **3.** For a subtle combination, the same allium is backed by the maroon stems of a dark-leaved penstemon. **4.** I would build a late spring bed around this gorgeous magenta byzantine gladiolus, with white alliums and purple perennials. **5.** A cool combination of blanda anemone with *Tulipa humilis* var. *pulchella* Albocaerulea Oculata Group, whose name is longer than the plant is tall. **6.** This tiny, bright, red-orange coral lily looks fabulous with a background of white, blue, and red annuals.

schemes. Yellow and blue, sometimes with a touch of white, is a classic spring combination.

Probably the most famous patch of muscari is the river of blue at the Dutch bulb garden at Keukenhof, outside Amsterdam. Thousands of blue muscari are planted to weave like a water feature through the trees, surrounded by yellow 'Dutch Master' daffodils. It is a very impressive sight. Copy this look on a small scale in your garden by planting a drift of muscari between some daffodils or combining yellow and blue bulbs in a container.

If you add brilliant red tulips or the vibrant red coronaria anemone to this mix, you could create a planting that is akin to a primary color paintbox. This combination of red, yellow, and blue is one of the brightest and most stunning of all. To continue this color scheme into the summer, combine yellow lilies, dahlias, or gladioli with bright red crocosmias, red cannas, and blue agapanthus.

If you are a fan of dark colors for a summer bulb color scheme, there are plenty of dark-leaved colocasias like 'Black Coral' or 'Diamond Head' and cannas like 'Australia' with their near-black foliage and red flowers. The best dark flowers for late in the gardening season are found in the rich and velvety colors of dahlias. Choose 'Arabian Nights' or 'Hollyhill Black Beauty' for deep, dark red-mahogany blooms.

If choosing bulbs by color is not interesting to you, then take the "anything goes" way of color planning. There are delightful bulb plantings that resemble a very colorful bowl of candies. The plantings are bright, jolly, and exciting. This is something that works with tulips. There are so many color choices that sometimes you may want to try them all and opt for a mixed bag. The medley of colors can be especially attractive when growing up through a green groundcover which provides the antidote to color overload. Later in the year, the many colors of dahlias can give you a similar cheerful effect.

A way to amend the color balance after planting is by picking any blooms that you feel clash with the rest of the mix. These can be used as

PREVIOUS: This iconic planting of a river of blue muscaris at the Keukenhof gardens in the Netherlands is brightened by the addition of white muscaris and banks of yellow daffodils.

ABOVE: The primary-colored spring-blooming flowers in this cottage garden stand out against the bright green leaves. Bulbs include red and yellow tulips, blue muscaris, and English bluebells.

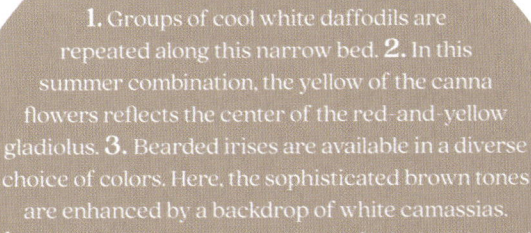

1. Groups of cool white daffodils are repeated along this narrow bed. **2.** In this summer combination, the yellow of the canna flowers reflects the center of the red-and-yellow gladiolus. **3.** Bearded irises are available in a diverse choice of colors. Here, the sophisticated brown tones are enhanced by a backdrop of white camassias. **4.** The narrow red-orange stripes of this dark-leaved canna pick up the colors in this single dahlia. **5.** A red-and-white-flamed tulip is underplanted with orange wallflowers (*Erysimum*). **6.** One flower of 'Peaches-n-Cream' dahlia contains a whole color scheme.

cut flowers in the house. Mixed bags of daffodils usually do not work as well as choosing the ones that you really want. The same with mixed bags of lilies, unless you want to grow them in pots and can rearrange them or use them as cut flowers.

Bulbs for a Fragrant Garden

Fragrance, like color, is an intensely personal matter. Fragrant flowering bulbs have characteristic scents that become indicators of a time of year: hyacinths are the epitome of spring aroma, and the fragrance of lilies tells me it is summer.

Some people are finely attuned to scents, while others cannot smell things very well. If you are a fan of flower fragrance, read the bulb descriptions carefully to check whether the scent sounds appealing. Investigate your preferences by smelling as many flowers as possible. In this way, you will find the ones that you want to add to your fragrant landscape.

There are many design ideas that will increase your enjoyment of scented bulbs. These include tucking your fragrant bulbs into a sunny sheltered corner to amplify the scent, planting them near a seating area, and raising the planting beds. Another thought is to grow them in pots on tables so that it is easy to enjoy their fragrance.

Scented Spring Bulbs

Spring bulbs emit a characteristic fresh, greenish smell that pervades the air. The first fragrant bulb of the year is the snowdrop. *Galanthus nivalis* 'S. Arnott' is best known for its scent of honey. Crocuses, cyclamen, muscaris, and rock garden irises may be fragrant too. Look for the dark purple scented rock garden iris called 'George'.

The most well-known of the fragrant spring flowers are hyacinths, but some daffodils and tulips have a scent too. Some of the top daffodils that are known for their scent include the jonquils, tazettas, and poeticus. For heavily scented tulips, try the old-fashioned 'Duc van Tol' or some of the parrot tulips. English bluebells have a delightful scent. The aroma is stronger and easier to smell when they are planted in a group. Later in the spring, bearded irises are slightly grape-scented.

Scented Summer Bulbs

In the summer, lilies are the star of the fragrant garden. Only certain lilies are scented, so read the descriptions before you buy. Trumpet and hybrid Orienpet lilies are renowned for their perfume. The regal lily has a super fragrance in early summer. For later in the summer, choose 'Stargazer' or 'Casa Blanca'.

Tuberoses are a classic scent of late summer. These make good cut flowers—just one or two will provide enough fragrance in an arrangement. *Hymenocallis* and *Galtonia* produce sweet summery smells. As summer turns into fall, acidanthera gladiolus has a fresh powder scent.

LEFT: In the summer, my porch is perfumed by 'Scheherazade' lilies leaning over the garden gate.

TOP RIGHT: 'Lady Derby' hyacinth is a historic cultivar that has been prized for its scent since 1883.

BOTTOM RIGHT: Tuberoses are delightfully scented summer-flowering bulbs. When cut and brought inside, one stem is all that you need to fill a room with fragrance.

1

2

4

1. A good place to grow flowers for cutting, like these purple tulips, is at the end of a vegetable bed. **2.** Dutch irises, such as 'Mystic Beauty', make a great addition to late spring bouquets. **3.** Asiatic lilies are easy to grow as cut flowers. There is a lovely range of colors available. **4.** Keep on cutting your dahlia flowers and more will be produced.

3

Growing Bulbs for Cut Flowers

Many bulbs produce flowers that are excellent for cutting. It is such a treat to be able to go into your garden and pick a posy of bulb flowers to brighten up your home. An excellent place for growing a few cut flowers is in your vegetable bed, where you can grow each type in its own row or block. Most gardeners do not have a separate cutting area, instead they wander through their whole garden and pick a few flowers here and there. If you want your bulbs to come back next year, make sure that you only cut off the flowering stem, leaving as much of the foliage as possible to renew the plant for next year.

In spring, there are so many bulb flowers to pick that you are spoiled for choice. Daffodils and tulips make excellent cut flowers. They can be picked when the bud is just cracking open and will continue to open in the vase. Small bulbs like scilla and muscari make exquisite mini posies. For late spring, choose Dutch irises, coronaria anemones, or ranunculus.

Summer and autumn bring lilies, gladioli, and dahlias to cut. The sticky, staining pollen of some lilies can be a problem; either place your arrangement on a hard surface or snip off the anthers. There are some lilies that have been bred for arranging that do not have anthers, such as the cultivar 'Corsage'. Gladioli are easy to grow and dramatic in a vase. To extend the cutting season, plant the bulbs in batches every week or two.

Dahlias are the darling of late summer and fall arrangements. They are super producers that can furnish plenty of blooms for yourself, or to give away. Dahlias need to be cut when the flowers are fully developed because they will not continue to open after being cut. Choose your dahlia plants by size of bloom, color, flower shape, and plant height. The taller dahlia plants take longer to come into bloom. However, once they begin producing, the large plants make many flowers.

Setting Your Bulbs Up for Success

Each type of garden bulb grows best in certain growing conditions. However, there are some general things that every bulb requires. All bulbs require water to initiate and maintain growth. They also all require much less water while they are not actively growing in the dormant phase of their lifecycle. The volume and the timing of water needed in the soil, during active growth, differs from bulb to bulb. There are some specialized bulbs that do well in moist soil, and others that need an extremely free-draining soil with a summer baking from the sun.

The amount of sunshine that a specific bulb needs also varies from bulb to bulb. Every bulb needs some sunlight energy when the shoot emerges through the soil. Certain bulbs are adapted to shady places, and others thrive in full sun. A position in full sun with free-draining soil suits the vast majority of bulbs.

While you cannot change the climate conditions or the weather patterns that hit your garden, you can alter the specific growing conditions in a flower bed to set the bulbs up for success. Do this by finding little areas of your garden (remember when we talked about microclimates?) that suit the needs of each bulb. Easy changes to make include amending the soil, raising or lowering the bed height, or altering the amount of sun on a bed.

Considering Sun and Shade

One of the parameters that will guide your choice of bulbs is the amount of sun or shade that your garden area receives. Most bulbs require six or more hours of sunlight. This is known as full sun. The beds in which these bulbs are planted should be open to the sky, unblocked by neighboring buildings or other structures. It is the same sort of area where you would grow tomatoes, sun-loving vegetables, and summer annuals, such as marigolds and zinnias.

Shade-Loving Bulbs

Areas that receive between two to six hours of sunlight are just right for shade-loving bulbs. These bulbs come from forests, woodlands, and jungles around the world and are naturally adapted to growing in low light levels. If you have a shady spot in your garden that you want to plant with bulbs, choose caladiums, colocasias, alocasias, and begonias for summer. In the spring, if your garden is in the shade of deciduous trees, you can grow scillas, erythroniums, winter aconites, snowdrops, cyclamen, leucojums, English bluebells, camassias, and a few lilies such as martagons.

When designing woodland- or shrub-backed plantings, choose bulbs that will multiply and perennialize. Keep the bulb groupings informal and find spaces between tree roots. Pull the planting areas away from the woody plant. Then, if your plantings are successful, the bulbs will seed into the soil closer to the trunk.

Creating Bulb Lawns

Planting bulbs in grass is a lovely way to incorporate additional flowers into your landscape. Wild-looking spring bulbs look wonderful when they are grown in irregular clusters, as if they had seeded in place naturally. Dutch gardeners refer to this lovely spring method of bulbs in grass as *stinzenplanten*, or stinze planting.

When planting bulbs in turf, look for an area of lawn grass that is quite thin. To create a long-term planting, bulbs and turf need to coexist in the same space. Choose a partially shaded, grassy area that is not watered or fertilized, particularly during the summer. Do not use weedkillers on the grass.

A lawn area that is partially shaded by deciduous trees is a great choice. Spring bulbs look especially lovely when paired with early-flowering trees like ornamental cherries, crabapples, amelanchiers, or magnolias that flower at the same time as the bulbs.

Leave this lawn un-mowed for six to eight weeks after the last bulbs have flowered. Delay cutting the grass and the bulb foliage to provide time for the bulbs to regenerate for next year. In this crucial period, bulb foliage has time to die down naturally. If you want to naturalize flowers, choose bulbs that are frequented by pollinators. These flowers will set seeds that eventually become new bulbs. The resulting relaxed look is better than any gardener could produce.

Mowing the grass surrounding your bulb lawn makes boundaries that give the whole scene an intentional look. You could mow a cut-grass path through the patch to allow access to your bulbs. I have one area of bulbs in grass that is defined by a raindrop-shaped perimeter pathway. Your lawn could also be edged by surrounding the

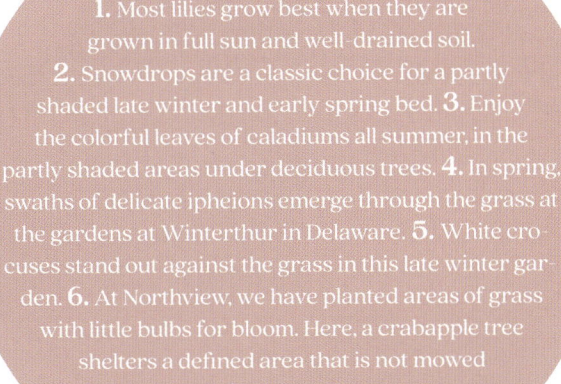

1. Most lilies grow best when they are grown in full sun and well-drained soil. **2.** Snowdrops are a classic choice for a partly shaded late winter and early spring bed. **3.** Enjoy the colorful leaves of caladiums all summer, in the partly shaded areas under deciduous trees. **4.** In spring, swaths of delicate ipheions emerge through the grass at the gardens at Winterthur in Delaware. **5.** White crocuses stand out against the grass in this late winter garden. **6.** At Northview, we have planted areas of grass with little bulbs for bloom. Here, a crabapple tree shelters a defined area that is not mowed until the summer.

SPRING- BLOOMING WOODLAND BULBS

The spring woodland in North America has an exciting group of shade-loving bulbous plants which are perfectly suited to growing beneath deciduous trees and shrubs. They capitalize on the spring sunshine to grow quickly from their storage roots. They need to have at least six weeks of sun to grow, flower, set seed, and send energy back to their root structure. They wait out the shady summer and the cold winter belowground before popping up again the following spring.

Some of my favorites in this category are bloodroot (*Sanguinaria*), Dutchman's breeches (*Dicentra*), merrybells (*Uvularia*), Solomon's seal (*Polygonatum*), trillium (*Trillium*), and Virginia bluebells (*Mertensia*).

ABOVE: Early spring brings fresh blooms to the deciduous woodlands of eastern North America. This is the pristine white flower of the bloodroot (*Sanguinaria canadensis*).

OPPOSITE: Virginia bluebells (*Mertensia virginica*) are spring ephemerals that grow from carrot-like roots.

1. Single daffodils are a perfect choice for naturalistic grass plantings. **2.** The well-known grassy meadows at Great Dixter in Sussex, England, have many flowers that pop up in sequence throughout the spring into early summer. The area is not mowed until the flowers have set seed. **3.** Autumn-blooming colchicums are strong growers that can compete with grass, as long as fertilizer is not applied to the area. **4.** Blue camassias are planted in large drifts to border the creek at Chanticleer, Wayne, Pennsylvania. This low-lying area of the garden is perfectly suited to growing wet-tolerant bulbs.

area with some strategically placed logs, stones, a small fence, or some shrubs.

Crocuses are one of the best bulbs for lawn plantings, especially tommies (*Crocus tommasinianus*). Other small-scale bulbs to consider in turf include early-blooming snowdrops, scillas, ipheions, and muscari.

Daffodils lend themselves to effective planting in weak grass. Look for daffodils that set viable seed like the poeticus types, and ones with a single graceful flower. The traditional choices for lawns have been the pheasant's eye daffodil and the short hoop petticoat daffodil.

Later-flowering bulbs that have a wispy look and long, thin stems, can also be used in grass. Try byzantine gladiolus, tall species tulips, the slender drumstick allium, camassias (in blue, purply blue, or cream), and autumn-flowering colchicums. Each of these lends a different look to your bulb lawn. Do not mow your summer-blooming lawn until late in the summer.

Bulbs for Wet Soil

Not all bulbs need well-drained soil to grow well. A small group of bulbs grows in wet soil for at least part of the year. They are a specialized bunch that have structural adaptations within the bulbs to resist rotting. In their native habitats, they are found in boggy areas around ponds or in fields that border rivers. Plant these ones in the soggy, puddly areas of your garden, such as near a downspout, in a rain garden, or around a regularly refilled birdbath. These areas do not have to be constantly moist so long as they are occasionally wet after heavy rains, especially during the growing season.

Checkered fritillaries are a good choice for planting near water because they grow well in seasonally moist soil.

The first wet-site-tolerant bulbs to bloom are the white-flowered leucojums and the checkered fritillary, *Fritillaria meleagris*. Later in the spring, the western North American camassias grow well in wet places. Summer bulbs that are water-loving come from warm or hot tropical and subtropical regions of the world. For wet soil, use tender alocasias, cannas, crinum, and calla lilies. Most bulbs need a period of wet to initiate growth and flowering, but some will only flower after rain, such as *Zephyranthes*, the rain lilies. In their dormant period, they require dry soils.

Raised Beds for Increased Soil Drainage

Most garden bulbs need free-draining soil to do well. One of the easiest ways to improve soil drainage is to elevate the level of the planting bed above the regular soil. If your garden is sloped, you may have naturally occurring raised areas

that will make excellent places for bulb planting. If you live on a flat or gently sloping site, you might want to create some artificially raised areas for your bulbs.

Raised beds can be an attractive design feature in your garden. The simplest way is to make a free-standing pile of soil of any shape. In most cases, these beds are edged with stone or wood to retain the soil. Use milled lumber to outline rectilinear beds and choose natural local materials to create an informal outline. Edged beds are often elongated in shape, which allows easy access to all plants.

From a design perspective, flowers in raised beds are easy to see as you walk alongside them. The beds provide a sense of enclosure for a path or a sitting area. Each side of a raised bed is subjected to different sun, wind, and water conditions. The sunny side of the bed is hottest and driest, and the shady side is cooler and wetter. The side facing the prevailing wind dries out first. The interesting thing about this is you can plant the same bulbs on the different faces of the raised bed, and they will flower in succession. By planting your bulbs in these different places, you are providing mini microclimates that vary slightly in growing conditions while only being a few feet apart. Part of the fun of growing bulbs is finding these limited areas of your garden that offer exactly the right conditions for each bulb that you want to grow.

Growing Bulbs in a Raised Rocky Area

When you are designing your bulb plantings, you might want to include an area that is rocky and raised higher than ground level, to mimic the mountain habitat of many bulbs. The presence of stones, rocks, or gravel, and the high bed

ABOVE: Raised beds, such as this narrow stone-edged one, have great drainage, suiting the needs of tulips and drought-tolerant perennials.

OPPOSITE: This raised bed at Northview is perfect for spring bulbs.

facilitate the movement of water away from the bulbs. Simply raising a flower bed an inch or two has some benefits for small bulbs. Larger bulbs need a taller bed so they can be planted at the right depth but are still above the rest of the garden soil.

Ideally, when you are designing this type of raised bed, you will be able to find some local stone that looks natural in your garden. Use large stones or rocks at the perimeter of the bed to retain the soil. Other rocks can be partially buried into the soil surface to mimic how they would be in nature. Rocks provide an excellent backdrop to showcase small and delicate bulbs. The top of the bed can be mulched with river gravel, or small chipped stones, to act as an aesthetically pleasing finish. This approach keeps the flowers looking pristine because it prevents mud from splashing onto the petals.

The spaces between large rocks become well-drained nooks where you can plant your bulbs. The bulbs stay dry, but the roots can grow down to find moist areas beneath the stones. Rocks provide stability and anchorage in the raised bed, so little bulbs do not get washed away in heavy rainstorms. Heat is absorbed by the rocks, which encourages the early growth of bulb flowers in spring, and keeps bulbs dry in summer. Planting in the shadow or shelter of a large rock provides a perfect spot to protect tiny bulbs from excess heat, wind, or desiccation. One other thing that I love about stable rocks is I can use them as stepping stones, which preserves the air spaces in the rest of the soil. Rocks can also be used as markers to be able to find your tiny bulbs in the spring. You could draw a sketch of the bed as you plant your bulbs, to show where the bulbs are in relation to the rocks.

I have found that rocky raised beds are a perfect place to grow miniature daffodils, rock garden irises, late-flowering alliums, and dwarf Formosa lilies. This is also where I grow some specialist bulbs that would not otherwise do well in my East Coast garden, with its erratic rainfall patterns. I have had success with several West Coast natives such as firecracker flower (*Dichelostemma*), mariposas (*Calochortus*), and triteleias. Other bulbs that have done well in these conditions are Juno irises, dwarf bearded irises, dry-loving fritillaries, and species tulips like *Tulipa bakeri* and *T. acuminata*. Beyond the practical benefits, the raised beds facilitate my admiration of tiny flowers, which would get overlooked in a regular plant-packed flower bed.

Using Gravel as Mulch

Another way to grow bulbs successfully is to make a gravel-mulched flower bed. A wide variety of bulbs thrive in these conditions. They can be interplanted with complementary herbaceous plants that grow well in dry soil, which is low in nutrients. Match the size of the gravel to the needs of the bulbs you are planting. Little bulbs need small-diameter gravel; big bulbs can still emerge through larger stones.

The bulbs get plenty of growing benefits from the rocky soil covering. The gravel protects the surface from losing water, so bulbs stay moist beneath the soil. However, the bulbs do not become waterlogged because the stones will not hold onto excess rain. Gravel absorbs heat, so slightly tender bulbs that need a warm climate may survive your winter temperatures. Aesthetically, the color of the stone is a lovely canvas against which many small bulbs really stand out.

I have had my gravel gardens for decades. During this time, the plantings have grown and evolved, but the basic principles have remained the same. The soil berms that I created are mulched with two to three inches of small river gravel. The gravel insulates the soil from drying winds. The root zone is generally moist even after weeks of no rain. I do not need to water these gardens at all. There is winter moisture available for autumn-planted bulbs to start growth in spring. Then in summer, the whole garden dries out during their dormant period. Even summer- and autumn-blooming bulbs seem to find enough water to grow and flower.

1

2

3

1. A rock garden in full sun is a great place to plant bulbs like *Triteleia* 'Corrina' that need summer baking. This waterfall is purely decorative and does not contribute much moisture to the surrounding soil. **2.** Rock gardens are great places to showcase petite jewel-like combinations, like this cream-colored *Fritillaria pallidiflora* and the perennial *Silene caroliniana*. **3.** This early-flowering miniature daffodil 'Mite' is only about six inches tall. It looks wonderful tucked between rocks. **4.** This area of my garden, called the Sunset Garden, is topped with a couple of inches of gravel. It is a perfect place to grow bulbs. The color theme here is yellow, so the year begins with winter aconites, yellow crocuses, and *Iris danfordiae*. By late spring, when this photograph was taken, the color is carried on by 'Baby Moon' daffodils. In autumn, *Sternbergia* flowers continue the theme.

4

I do not use organic materials on areas where I have gravel mulch. If deciduous autumn leaves fall on these flower beds, they are cleared off. More stone is added to the soil surface as needed. Because the soil in these gravel-topped areas is not fertilized, it remains "lean and mean," with low nutrient levels. My gravel beds are not watered, so bulb plants grow slowly. I want the bulbs to grow short and stocky to avoid the need for staking.

Drought-tolerant bulbs are planted in the soil beneath the gravel. To do this, I push aside the gravel, plant the bulbs as normal, firm down the soil, and replace the gravel. This is my preferred area to grow lycoris, crocuses, colchicums, sternbergias, blackberry lilies, and some medium-sized daffodils like 'Rapture' and 'Jetfire'. If you garden in a warm area, try nerines, agapanthus, or society garlic in your gravel garden.

ABOVE: *Iris reticulata* 'Pixie' is an early spring bloomer in an area mulched with small pebbles.

OPPOSITE: *Lycoris incarnata,* peppermint spider lily, is one of many *Lycoris* that grows well in gravel-topped beds.

Adding bulbs in containers can bring color and excitement to any part of your garden. On this patio in my garden, pink and purple tulips, muscaris, scillas, and blanda anemones are mixed with a few cute violas.

4

THE BULB CONTAINER GARDEN

Bulbs and pots were made for each other. Most bulbs need excellent drainage, and containers can easily provide that. Containers need to be showy and full of visual interest, and bulbs can provide that too.

Any garden, no matter the size, can benefit from having some bulbs in containers. If you garden solely in containers, you are going to love how much bulbs add to your display. For a garden that has in-ground beds, containers are a way to grow plants that might not do well in your soil. They bring the beauty of bulbs front and center. Spring flowers in pots add a punch of color to any area. Potted warm-loving bulbs provide delight in the summer and fall with showy flowers and bold leaves.

There are several practical reasons for growing bulbs in pots. First, the environment in a container allows you to choose the correct soil for the bulbs that you want to grow, and to determine how much water to provide. In addition, the pots are moveable, so you can put them in the right sun or shade conditions. Containers can also be taken with you when you move to another home, making them a great choice for renters. Raising up potted bulbs brings them into clear view. They can be placed at a comfortable height, so that anyone can tend the container garden.

Some gardeners like to grow bulbs in pots as annuals to provide the biggest and most extravagant floral display. As soon as the flowers start to fade, the bulbs can be removed and replaced. The loose soil in the pots means that switching out the displays seasonally is easy without having to dig in garden soil.

Choosing bulbs to plant up in pots is a lot of fun. Think of each container as a miniature display that decorates your garden for the season, and is then replanted with different bulbs for a later show. To personalize your bulb garden, select by color, fragrance, or showcase a collection of your favorites.

Situating Your Bulb Pots

The best place to put containers is where you spend time in the garden and will see them regularly. If the pots are close at hand, you will be able to admire them. When putting together a container display, consider the bulbs and pots that you are combining. Think of the proportions, plant height, flower shape, and color. Unless the pot itself is the focal point, choose a pot color and texture that blends into the background, allowing the flowering bulbs to shine.

In large gardens, bulbs in pots are used to add color and seasonal interest around living spaces, and as accents in the rest of the garden. A pair of bulb-filled pots can flank a path or act as a visual gateway from one garden area to the next. Potted bulbs can also be used within, or next to, your flower bed. These pots can be elevated on a large flat rock or a pedestal for a dramatic look. If you plant multiple containers with a consistent planting scheme, they add cohesiveness to the garden. To increase visual harmony, arrange them along a path, around a patio, or within a flower bed. In small gardens, balconies, and roof gardens, all the bulbs may be in pots. Regardless of the size of your garden, creating vignettes using multiple pots is a great way to design with bulbs.

Make a Vignette of Potted Bulbs

Container vignettes are displays made up of a collection of potted bulbs, clustered together. A pleasurable part of container gardening is that you can personalize your display by selecting which pots you like, the bulbs to go in them, and how you want to arrange them. Think of your container collection as a fluid composition that changes as the bulbs go in and out of flower. You can bring the pots that are having their prime time to the front, while shifting the ones to the back that still look good but are less attractive.

The pots could be at ground level, raised up on tiered planks, or on a tabletop. Raising the containers off the ground has aesthetic benefits. You can see the contents more easily than if they were on the ground. Place the potted bulbs so that the flowers are at eye level when you sit down. Raising the containers also helps with drainage. And clustering them makes watering more efficient.

When putting your vignette together, begin with your largest pot. Arrange other smaller pots around this focal-point pot. Add in other containers of annuals, perennials, or woody plants to provide additional textures and foliage colors. If you want the containers to stand up high, sit them on a pile of bricks or overturned empty pots.

Keep your vignette looking at its best by turning the pots around so their best face is to the front. Pull the best-looking pots forward and move others to the back. While you are admiring them, do a little deadheading or remove a faded leaf.

I like to arrange containers along my porch wall. When I am having breakfast, the little beauties are right there. I can watch the comings and goings of the bees and smell the fragrance that is attracting pollinators to the flowers.

1. This summer container collection uses bulbs such as cannas and tulbaghias as part of its colorful display. 2. A hanging basket, like this one filled with begonias, is a cheery addition outside a door. 3. Yellow, blue, and white bulbs fill containers that frame this bright blue door. Consider the colors of your house when deciding what bulbs to put in your pots.

1. Clustering pots of bulbs together, as on this tabletop, allows you to combine and recombine as you wish. Here, the double yellow 'Verona' tulip is behind the delicate white 'Xit' daffodil, orange 'Prinses Irene' tulip, light blue 'Baby's Breath' muscari, and the yet-to-open multiflowering 'Danique' tulip. **2.** The fall display on the same table features a different color palette. 'Candidum' caladium, with its white-and-green leaves, stands out against a background of eucomis, tuberoses, non-bulbous coleus, heuchera, and a few succulents. **3.** The bright pink geraniums and deep brick-colored petunias coordinate strikingly with the reddish flowers of a canna in this potted display. **4.** Early spring potted bulbs are delightful when observed up close. A pot of *Scilla bifolia* 'Rosea' has been placed on an elevated patio wall with a container of white *Puschkinia libanotica* 'Alba'.

Choosing the Right Containers

Choosing the right containers will help you successfully grow bulbs in pots. The overall look of the containers should complement where you want to place them, whether near the house or out in the garden. Some people prefer a cohesive set, while others may collect eclectic shapes and colors. Practically speaking, containers must suit the growing needs of your chosen bulbs. Things to consider include the pot's dimensions, weight, material, and whether it has drainage holes.

Pot Materials and Drainage

When selecting your bulb pots, the most important thing to look for is the presence of a large drainage hole or several smaller holes in the bottom. Most bulbs require excellent drainage. Planting into an undrained pot is rather like putting bulbs into a bathtub full of water where they will quickly rot and die. It is possible to have an outer decorative pot with no drainage holes into which you insert a practical inner pot that must have holes. Be sure to regularly empty standing water out of the decorative container.

Depending on your climate, the material that the pots are made from may affect how well they serve your bulb plantings. In moderate climates, you can use any type of pot. In hot climates, steer away from dark colors or metal containers, which heat up quickly. In cold climates, check whether your pot will be able to withstand your coldest temperatures and the terrible winter combination of cold and wet. If you are unsure how well your pot will do in cold weather, bring it inside for the winter.

Wood is an excellent material for window boxes. It is great for bulbs because it doesn't heat up in the summer and will drain well. Here, white *Begonia boliviensis* 'Santa Barbara' is mixed with white and blue annuals that pick up on the garden shed's colors.

Plastics and resins can distort or break but they make good short-term bulb pots because they are usually cheap to purchase. They are light in weight and easy to move. One downside for bulb culture is that they are impermeable to airflow and can hold too much water. Summer bulbs that like wet soil, such as cannas, do well in these pots.

Metal, concrete, stone, and wood will last in all but the toughest conditions. Metal, especially if dark in color, can get too hot for summer bulbs. Concrete and stone are extremely heavy to move, so place them carefully in their final positions before adding the bulbs. Wooden half barrels or window boxes are good natural-looking choices for bulb growing. Wood is also beneficial for bulb containers because it does not heat up in the summer. Drill holes in the bottom of wooden containers to ensure that they drain efficiently. Wood is quite frost resistant, so it is a good choice for winter and early spring bulb displays that stay outside all year. I have had some old wooden window boxes and containers. They still look good, but over time they will gradually disintegrate and will need to be replaced.

Terracotta, a type of unglazed earthenware, is a traditional choice for growing bulbs. This material allows for excellent drainage and airflow. It ranges in color from browns, beiges, and rusty oranges, depending on the clay from which the pots are made. Choose your bulbs to complement the pot color. If left outside, these pots are prone to crack or flake in winter. There are pots that are labeled "frost resistant" or "frost proof." Frost-resistant pots offer a slight resilience to winter damage; frost-proof pots are even tougher, though not guaranteed against cracking.

Glazed pottery is not usually left outside in winter because it can also crack. Glazed pots are widely used in gardens because of the variety of colors that are available. They are wonderful for personalizing your displays. I am a fan of bright blue glazed pots as they work well with a variety of different bulb displays. Pots in neutral hues, such as black, gray, or dark olive greens are other good choices that recede into the background, allowing you to choose any bulb color that you like.

All pots benefit from being stored in protected areas over winter, whether they are empty or full of bulbs. For large pots that are a permanent fixture in your garden but not planted up, empty out the soil and securely cover the top with a waterproof fabric. It is not necessarily the prettiest winter look, but if it saves a precious pot, it is a good thing.

Pot Size and Volume

The height, width, and total soil volume of the pot are important factors to consider when choosing a container for your bulbs. Your pot needs to be wide enough to accommodate as many bulbs as possible to make a good display. The depth is important to accommodate root growth, for absorbing water and holding the bulbs vertically in the pot. The total volume needs to be large enough so that the soil does not dry out quickly and desiccate the bulbs.

Match the size of the pot to the types of bulbs that you want to grow. Large bulbs usually develop into correspondingly tall plants, which need to be planted into deep pots. These containers provide plenty of room for their roots. Large bulbs also need weighty containers so that they do not topple over when their potentially top-heavy flowers are in bloom.

Large pots are useful if you garden on a windy site such as a rooftop, balcony, hillside, or beach-front. Look for pots with stable bases to help the pot stand up. If necessary, fix containers in place to lessen the risk of them blowing over or add heavy stones into the bottom before planting. Check weight restrictions before placing heavy pots on decks, balconies, or rooftops. Include the weight of soil, stones, and plants in your calculations. Pots with large volumes dry out more slowly than small pots.

When you are creating mixed bulb arrangements, choose the biggest pot that you can find. There will be lots of bulbs and many roots competing for water and nutrients in the finite soil volume of a container. These giant pots are difficult to move once you have potted them up, so place them in their final positions from the start.

Terracotta containers can be whimsical in shape, as with this chicken pot that is filled with white scilla. Behind it, the daffodil 'Spring Sunshine' is growing in a plainer terracotta pot.

Shallow and wide containers, such as those called bulb pans, look just right for miniature bulb flowers like crocuses, snowdrops, and little rock garden irises. Little pots have lots of surface area compared to their volume. They will need watering more frequently than large ones under the same weather and watering regime. We have found that it is difficult to keep plants moist, but not soaking wet, in containers under eight inches wide and tall.

To achieve a pleasing overall look of the finished container, check the proportions and heights of both the plants and the pots. The pots should frame, but not overwhelm, your plants. To look balanced, the plants and the pots should be either about the same height as each other, or the overall ratio should be two-thirds plant height and one-third pot.

When buying pots, make sure that the mouth of the pot is wide enough to allow you to change out the bulbs easily. I have bought pots with narrow necks that I loved for their looks, but that were impractically shaped. The easiest pot shapes to use for bulb growing are those that resemble buckets used to make sandcastles at the beach. You can easily turn them upside down, empty out the contents, and sort out the bulbs before replanting.

Choosing and Combining Your Bulbs

Once you have your containers, you can select your bulbs. Begin with your favorites and choose others that bloom at the same time to create a floriferous pot. Consider heights, colors, flower shapes, plant form, and foliage color. Whatever bulbs you put in one pot should all need the same growing conditions.

The easiest way to plant up containers is to put one type of bulb in each pot. These one-bulb pots are both quick to plant up and simple to look after because the bulbs all require the same growing conditions. The matching bulbs bloom at the same time to produce a lovely flowery display. Plant up multiple single variety pots to make a flexible garden that you can rearrange however you like.

When preparing multiple bulb containers for a vignette, consider how your display will look together as it grows. If you want the pots to look cohesive, repetition of some aspect is important. Choose containers that are made of the same material or repeat the type of bulbs or their colors in the different pots. Add interest with a variety of flower shapes or by using plants with variegated foliage.

A slightly more complex way to grow bulbs is to mix several different types together in one large pot. The soil volume needs to be sufficient, however, so that all the bulbs will have space to grow. Make sure the bulbs that you put together all need the same growing conditions, including drainage, moisture, and sun or shade requirements. When they are in bloom, these delightful combination pots are eye-catching and impressive.

1. Small bulbs look great when planted in proportionally sized containers. **2.** This is the same pot five months later. The white flowers of *Scilla luciliae* 'Alba' stand out well against the dark gray container. **3.** These large pots have layers of plantings with the imperial fritillary 'Lutea Maxima' in the center surrounded by tulips—the yellow 'Jan van Nes' and the eye-catching 'Apeldoorn's Elite'. **4.** These leafy summer bulbs combine well with bright magenta-pink impatiens in part shade.

Planting Your Bulb Pots

Before planting up your bulb pots, lay out your supplies. You will need your chosen bulbs, containers, and plenty of potting soil, as well as grit, gravel, stones, or broken shards of old terracotta pots called crocks. It is often easier to pot up bulbs into containers on a sturdy bench or table, rather than on the ground. If the pot is heavy or immovable, you will need to plant it up in place.

No Soggy Bottom

When you plant up bulbs, give them the best possible chance of growing well. Maximize drainage and most bulbs will flower profusely. One of the main enemies of good bulb growth is excess water held in the soil that rots and kills bulbs. Keep drainage holes open so that water can flow out while potting soil stays in.

Add stones or crocks to the base of the pot and top this with a layer of fine grit or coarse sand. In some places, you can buy grit that is labeled for horticultural use. A good alternative is grit used for chicken husbandry. If you cannot find grit, substitute a coarse sand, small-scale gravel, or tiny pebbles. Buy stone or sand that comes from your local area, if possible, to keep the mileage for shipping low.

Potting Soil Mix for Bulb Containers

Growing bulbs in containers gives you the chance to customize the potting soil to suit the bulbs that you plant. The mixture described here is a good general one for most bulbous plants.

Select a potting mix that is labeled for container growing, preferably one that does not contain peat. Peat is an unsustainable resource that is harvested from peat bogs, which are excellent carbon sinks. Peat bogs also provide vital habitats for unique plants and animals. Alternative potting soil ingredients include coir, broken-down leaves, and other organic matter.

To improve drainage for bulbs, amend the potting soil. Add more of the grit that you used at the bottom of the pot. I use about a third grit by volume. Using an old container as a measuring scoop, I add two scoops of potting mix to one scoop of the grit in a wheelbarrow or big bucket. With a shovel, combine the ingredients to produce a soil that has an open texture. Bulbs need air in the soil to allow drainage and to encourage good root growth. Moisten this mix lightly with water and stir again. It should be moist but not wet.

You can change the proportions of grit and potting soil according to the type of bulb that you are planting. If you are growing large bulbs in containers as a permanent installation, you may also want to add some ordinary garden soil to the mix. This adds weight and structure to the combined soil. If it is a clay-based soil, it will help to provide nutrients.

Certain lilies need a specific acidity or alkalinity to do well. Some large summer-blooming bulbs, particularly those like cannas and alocasias, from tropical and subtropical areas, need a large soil volume and plenty of water. Their potting soil should have extra compost, a small amount of slow-release fertilizer granules, and maybe some water-retaining additives. Follow the instructions and be careful not to use too

1. Agapanthus bulbs thrive packed closely together in this terracotta pot. They are clustered with containers of drought-tolerant plants on this terrace wall at the gardens of Chanticleer in Pennsylvania. **2.** Lilies are great in pots but they need excellent drainage. This one is called 'Garden Party' and is about two feet tall, so it is a good choice for containers. **3.** For a neat look to the finished pot, and to also stop potting soil from splashing onto flowers, top-dress with sand or gravel. These dramatic bicolored flowers are 'Spitsbergen' tulips. **4.** When planting bulbs in containers, the depth and spacing often differs from in-ground placement. These hyacinth bulbs are placed close together, but not touching, to create a full look when they are in flower.

much. Never use water-retaining gels for bulbs that need good drainage.

Planting Depth for Containers

Each type of bulb has a recommended planting depth. If in doubt, measure the height of the bulb. Place the base of the bulb in a hole that is two to three times that depth. There are some exceptions for growing certain bulbs in pots. Nerines, amaryllis, and crinum should be planted with the necks of the bulbs poking out of the potting soil. Cyclamen and tuberous begonias should be planted just below the soil surface.

Bulb Spacing in a Container

Place the bulbs so that they are close together without touching. Ignore the stated spacing distances on the packet because those usually relate to in-ground planting. To produce a beautiful full-looking show, your bulbs should be packed together in the container. Plant the bulbs with their pointed tips upward. If in doubt about their top or bottom, lay them on their sides. When planting large bulbs like daffodils, hyacinths, tulips, begonias, or caladiums, odd numbers look best in a pot. Try three bulbs arranged as if at the points of a triangle, or five positioned like the spots on a die.

Cover the bulbs with additional grit-enhanced potting soil to fill the container to within an inch of the rim, or half an inch for shallow pots. Gently press the soil to make sure that there are no big air pockets around the bulbs. This gives you enough room to gently flood the top of the pot with water, letting it slowly infiltrate throughout the pot. All bulbs need to be hydrated evenly. Any excess moisture should come out of the base. When water seeps out of the drainage holes, you know that you have watered the pot sufficiently. Finally, top the soil with a layer of grit or gravel as a dressing. This gives the surface a finished look and prevents the brown potting soil from splashing onto the flower petals.

Seasonal Displays in Containers

One of the best things about growing bulbs in pots is you can do some fun seasonal displays with different bulb combinations for every part of the gardening year. The spring display will be planted in your pots in the autumn and the summer through fall display will be put together in late spring.

Spring Is a Long Season

Spring is a long season that can be broken down into early, middle, and late. If you have several containers, you could plant up an early, a mid, and a late spring pot, rotating the one that is in prime condition to the front of the display. You could also combine a variety of bulbs that bloom at different times of spring into one pot by using what is known as the lasagna method. This bulb technique is where you plant bulbs at

These large, deep pots are packed with spring-flowering bulbs to give a full show. The front pot contains 'Pipit' daffodils, with tulips, including 'Pink Sound' and 'Foxtrot', in other pots behind it.

different levels in the pot. It gets its name from the well-known layered pasta dish.

To create a floriferous spring pot, you need to pack in as many bulbs as possible. They should be planted much closer together than if they were planted in the ground, and in several layers. The biggest bulbs go toward the bottom of the pot, then the medium-sized ones, with a final layer of tiny bulbs on top. The little bulbs will usually flower first, followed by the medium-sized ones. The large

bulbs are often the last to bloom because they take a while to grow up through the other levels.

When planting a layered pot, choose a deep container. Put crocks, grit, and free-draining potting soil into the pot first. This is where your big bulbs will grow their roots, so provide at least a few inches of soil mixture at the bottom. Sit the largest bulbs on this base layer. Cover them with a layer of potting soil mixture. Feel where the lowest bulbs are located and nestle the next layer of bulbs in the gaps between the bottom ones. Continue with another layer of the potting mix and the final layer of smaller bulbs. Top with the remaining soil and grit. The large bulbs will easily come up between those above them.

Pots with a large soil volume might contain a bottom layer of imperial fritillaries, camassias, large alliums, tall daffodils, or tulips, topped with short ones. For a simple pairing, select your favorite daffodils, adding some small blue bulbs such as scillas or muscari around the edges.

If you stick with one color combination, it will not matter if one bulb blooms earlier or later than expected because they will all coordinate. When the flowers emerge, the resulting display will be full and colorful.

Containerized bulbs get much colder than bulbs that are planted directly into the ground. This is because they do not receive the insulating effects of being in the earth. Pick tough bulbs for your pots that are listed for your growing zone or a zone or two colder, depending on how exposed your pots will be in winter. If you are placing them on a windy rooftop or balcony, the conditions are similar to an exposed mountain. Luckily, many spring-flowering bulbs grow naturally in cold climates. If you have plenty of snow in the winter, this is an advantage. The snow acts as an insulating layer to keep your pots at a steady temperature.

ABOVE: Add little annuals, like this short viola, into your bulb pots in the spring.

OPPOSITE: A collection of containerized plants looks coordinated in blues and whites on a garden wall against a background of early-flowering spireas. Bulbs include *Scilla siberica* 'Alba' and the light blue muscari 'Julia'.

COMBINING ANNUALS AND PERENNIALS WITH SPRING BULBS

To further increase the showiness and diversity of your pot, you can add other spring-flowering plants to the soil around the bulbs. In warm winter areas, you can plant the pot fully in autumn. If you have cold winter weather, then plant up the bulbs in autumn, moving the pot to a sheltered position until early spring. When you bring it out, you can buy some small additional cool-weather annuals to add fullness to your bulb pot. These can be bought in little packs with tiny root balls, which are perfect to tuck between the emerging green bulb noses.

The list of possible plants that you could use is long. Look for cool weather–tolerant flowers, such as pansies, English daisies, sweet alyssum, snapdragons, wallflowers, candytuft, stocks, trailing lobelia, forget-me-nots, nemesia, and primroses. Leafy greens like lettuce, hardy herbs, ornamental kales, or Swiss chard can also be used. Any of these could alternatively be potted up on their own to use when creating container vignettes. Pack the plants in tightly, so they weave together, mixing and mingling.

If spring rolls around and you realize you forgot to plant up your pots with bulbs in the autumn, do not fret. It might cost you a little more, but there are often potted bulbs for sale that are already growing, which you can assemble to make a great container. Look for small-scale daffodils, like the ever-popular 'Tête-à-Tête', hyacinths, or tulips. Combine with other cold-tolerant flowers.

BULBS THAT DO WELL IN SPRING CONTAINERS

Early spring brings a host of delightful small-scaled bulb flowers to the garden. If they are raised up in pots, small dangling flowers like snowdrops are easy to see. If you have some popping up in your garden, and the ground is not frozen, you can dig some up. Put them into a pot so that you can watch them bloom close up. Other good choices are rock garden irises and crocuses. If you garden on a heavy, wet clay soil, both of these bulb types are better grown raised up in pots. They need excellent drainage, so their potting medium should include grit, fine gravel, or sand, as half of the total volume. Crocuses are a favorite edible treat for many animals, so by growing them in pots, you can protect them. Later in spring, try muscari or scillas.

Spring would be a sad season without the exquisite fragrance of hyacinths. A pot of hyacinths can scent a whole courtyard or terrace. Daffodils are the most reliable bulbs that you can grow in pots. The diversity within this genus means that there is a daffodil that will suit any container. Match the height of the daffodil and the scale of its flower to the size of the pot. For small pots, choose miniature or intermediate-sized blooms. Large pots can take full-sized daffodils.

Tulips look fabulous grown in containers. Their beautiful shapes and rainbow's worth of colors look wonderful when clustered together and raised for easy viewing. Growing tulips in a container gives them some protection against burrowing animals and browsing deer. Use an animal repellent in and on the pot. Try using decorative twigs around the tulips to defend them and to hold up the tulips in an artistic manner. For showy cultivars of tulips, you get the best display with newly purchased bulbs.

1. If you have snowdrops growing in the ground of your garden, you can dig some up and put them into a pot while they are in bloom. **2.** Hyacinths, like this violet-purple 'Miss Saigon', have an enchanting fragrance that is easier to smell when raised up. **3.** Muscari grow well in pots, including this lovely white to pale blue one. **4.** This short orangey tulip, *Tulipa praestans* 'Shogun', opens in the sun to reveal dark stamens.

1. If given sufficient space, summer bulbs, like this canna, make an impressive show. Here in Manito Park, Spokane, Washington, a canna is underplanted with sweet potato vine, petunias, calibrachoa, and dichondra. **2.** Calla lilies bring a different flower shape in a range of colors to your summer containers. **3.** Short dahlias make excellent choices for summer pots. When selecting a bulb cultivar, look at the colors, sizes, and shapes of your containers. **4.** In climates where tender bulbs are not hardy, it is sometimes easier to grow them in pots. These eucomis bulbs do really well in containers.

If you want to reuse the container for the summer, remove the bulbs and their attached leaves. Hardy spring-flowering bulbs can be immediately replanted elsewhere in the garden, where they may bloom next year. They can go straight into a flower bed, into an area of rough grass, or they can be replanted in an out-of-the-way place, like the back of a vegetable garden. Small bulbs can be removed from containers and tucked into the front of a flower bed, where they will be a nice surprise next spring. If you no longer want the bulbs, they can be composted.

Summer Through Autumn

There is a different set of bulbs that become the stars of your summer and fall containers. Most of these summer bulbs need a large soil volume to accommodate their extensive growth. In the right conditions, they can grow rapidly to make an impressive display. Your summer bulbs could be grown each in their own pot or combined into a mixed pot.

To get a jump-start on the growing process, you can start them inside, transferring them to the display pot when nighttime temperatures are consistently warm. Before planting outside, they will need to be hardened off. This means that you gradually acclimatize the bulbs to the outdoor conditions. To do this, move the bulbs out during the day and back in at night, when it is cold. Follow this routine for about a week. Later in the season, you may be able to find potted bulbs that are already in growth. These can be replanted into decorative containers and added to your bulb collection.

The bulbs of summer bloomers vary tremendously in size, shape, and planting requirements. When planting your pots, look at the depth of planting and spacing each bulb needs. Match the size of your pots to your chosen cultivars. If your pots are moderate in size, look for smaller types.

As you assemble the bulbs for your summer pots, check their growing requirements and combine those that thrive in the same conditions. Some bulbs can be grown in full sun, while others do best in part shade. Depending on the intensity of your summer sun, the same potted bulb may need more or less sun exposure. For example, begonias may need part shade in hot zones, but may benefit from full sun in other areas.

Heat is also a factor that can affect bulbs in containers, more than those planted in the ground. If you garden in a cool summer climate, look for a sheltered nook near a sunny wall. The radiant heat coming off the wall will keep the containers warm. In hot summer climates, potted plants can heat up too much, especially in the afternoon, so look for a position in partial shade.

BULBS FOR SUMMER CONTAINERS

For the plants to do well, summer bulb containers often need extra water and fertilizer. Plants like cannas and colocasias do best with consistently moist soil. Dahlias are heavy feeders and require regular watering. Gladioli need well-draining soil and additions of fertilizer.

If you want to make vignettes of potted bulbs, you could plant up each summer specimen individually. Cluster them according to their sun or shade needs. Summer bulbs produce lots of bright hot colors and bold foliage. Think about both the flowers and the leaves as you create coordinating color schemes or brilliant contrasts. Some of the easiest bulbs to grow in containers are begonias, gloriosa lilies, caladiums, rain lilies, short dahlias, and lilies. Tall plants may need to

ABOVE: Rain lilies, like this white-flowered *Zephyranthes candida*, are fun to nestle among other summer bloomers, such as salvias and echinaceas.

OPPOSITE: Cute lilies like 'Corsage' look good in combination with other potted plants that pick up their colors, such as these heucheras.

be staked or surrounded by plant supports, like hoops or obelisks.

Planting lily bulbs in containers allows you to control the conditions and prevent rotting. Use a deep pot and loose, free-draining soil that promotes good root growth. Terracotta pots are my favorite material for lily pots because they are porous, letting air in and out to prevent the bulb from becoming waterlogged. If you grow a variety of lilies that bloom at different times, you can bring out each type, when it is nearly in flower, to include in your vignette. The resting lilies can be kept in a protected spot. Short-statured lilies are easy to grow in pots because they do not fall over. Large lilies will need proportionally bigger pots. Add a large stone or two to the bottom of the pot to counterbalance the weight of tall varieties.

If you love a tropical look, choose bulbs that produce large leaves and have vibrant colors. Include alocasias, hymenocallis, calla lilies, pineapple lilies, gloriosa lilies, caladium, and begonias. For a classic flower garden container collection, make a group of dahlias, gladioli, lilies, and liatris. You could choose agapanthus, tuberoses, and tulbaghias to line a warm terrace. To carry your display into autumn, add in pots of acidanthera, nerines, and hesperanthas.

THE MIXED SUMMER CONTAINER

Mixed summer containers can be lush and bountiful. Use large pots as many of these bulbs grow very fast and need lots of water and fertilizer. If the roots do not have enough room to grow, the plants will become stunted.

Summer bulbs often become the centerpiece of a large pot, especially large-flowered lilies, dahlias, and bulbs grown primarily for their leaves, such as cannas, colocasias, and alocasias. The shape and color of bulb foliage plays an important part in a successful mixed summer pot. Leaf colors may be green, bronze, near-black, or yellow. Leaf shapes may be elongated, broad, rounded, or heart-shaped. Elevating the pot allows you to admire the details of the flowers and leaves.

Decide on a feature plant, then play off the colors when choosing the other plants to add around it. The colors of both flowers and foliage play an important role in making a successful mixed summer bulb pot. Things to consider include whether you want to contrast or coordinate colors, the textures, and the plant habits. Include some plants to trail over the rim of the pot, and also ensure that you have adequate height. The mid-height plants in the pot should be chosen carefully. They are an important link

between the tall and the short, and should tie the whole pot together visually.

Once you have chosen the bulbs, look for other summer plants that you can add to the mix. There is a wide choice of companion plants that will bring in different textures and colors. Some of the best include warm-season flowering plants like marigolds, pentas, petunias, calibrachoas, geraniums, celosia, rudbeckia, and salvias, as well as foliage plants like coleus, sweet potato vine, and dichondra.

Changing Out Bulbs for Each Season

For an easy planting method that requires no digging, permanently sink one or more utilitarian pots into the potting soil of a large decorative container. Each seasonal group of bulbs is then grown in the same-sized plastic or terracotta pot, which can be switched out for each flowering time. After the first set of bulbs has finished flowering, take out the inner pot and replace it with bulbs for the next season. Planning ahead is important. You will need to be always thinking and growing the pots for future displays. When the old bulbs finish, you will have the new pot waiting to take its place.

For instance, you could begin the gardening year with pots of early daffodils that you swap out when they have finished blooming. Later on, use tulips, followed by Asiatic lilies, and then short dahlias to end the year. Each of these bold and beautiful bulbs can be a showpiece in a pot. Plant small annuals in the soil around the practical pots. These bushy or trailing plants will help hide the pot rims, while adding to the diversity of color and texture.

When setting up this system, think about the relative sizes of the outer and inner pots. A large urn or decorative pot can hold two, three, or more smaller pots, depending on its diameter. A smaller urn that fits just one changeable pot can also be used. This can be a good method for window boxes and other containers in prime locations, where you want them to always look great.

Watering Bulbs in Containers

When caring for bulbs in containers, watering is one of the most important things that you will do. The containers are totally dependent on you to take care of their water needs. Before you embark on bulb container growing, consider how many containers you can realistically look after. Place them so that they are close enough to your daily activities that you will notice when they need attention. Cluster them together to save time when watering and to minimize water use. Situating them near an outdoor tap makes this easy. The pot's material will influence how often you need to water. Large containers tend to need watering less often than small ones.

When deciding whether your potted bulbs need watering, feel the soil under the layer of surface grit with your finger. Water the pot if the soil is dry. There are two ways to water, either from below or from above. Very delicate little bulbs may be watered by sinking the whole pot into a bucket of water up to its rim, until the top of the soil feels moist. Then remove the pot from the water and let it drain.

1. Plastic pots of bulbs, like these yellow-and-orange daffodils, can be nested into larger decorative ones easily, and then removed to make way for the next show. **2.** This pot of 'Tiny Double Dutch' lilies is going to replace a spring-flowering bulb that has already finished. **3.** With a cascading habit, begonias are a good choice for a hanging basket. They will need regular watering because hanging baskets dry out quickly.

ABOVE: Tropical bulbs, like this alocasia, require regular watering and fertilizing to grow well.

OPPOSITE: Pot feet, here tucked under a container of 'Apricot Parrot' tulips, help the soil drain well when thoroughly watered.

If you are watering your bulbs from overhead, try to mimic gentle rainfall rather than a blasting firehose (which is something that I have been accused of doing). Use a watering can with a fine-holed end, called a rose, or the spray attachment on a hose. Add water all over the surface, so it sinks into the soil evenly. Water the soil and not the plants themselves. Continue to add water until you see it coming out of the bottom of the pot. This is your indicator that all of the potting soil is moist, and the roots of the bulbs are getting the water that they need.

If you are growing bulbs that need good drainage, do not let them sit in water for extended periods of time. You may want to have a saucer under the pot to protect a surface. If you do, empty the water out regularly. The same applies if you use decorative cache pots around your utilitarian pots.

To maximize drainage, use pot feet underneath your containers to prop them above ground level. This is especially useful for large- and medium-sized containers. Pot feet are small blocks that sit under the bottom edges of a container to raise it off the ground. This ensures that water flows out of the pot and air circulates beneath it. You will need three for a medium-sized container and five for a large one. Odd numbers of pot feet evenly spaced around the bottom work best, in the same way as a three-legged stool stands up steadily. In cold climates, pot feet prevent the pot from freezing to the pavement, or the soil beneath it, which decreases the chances of the container cracking.

Fertilizing Bulb Pots

Containers have limited soil volume, which means you need to provide the fertilizer they require for good growth. The rate and frequency of fertilizer application depends on pot size, bulb type, potting soil type, and weather. If you water your pots frequently or live in a rainy climate, fertilizer can quickly wash out of the bottom of the pot. If the soil volume is small and the plant is large, your bulbs will need extra fertilizing. The growth rate, leaf color, and flower size will inform you whether to fertilize or not. Indicators of subpar growth include weak stems, yellowing leaves, and small-sized blooms.

When it comes to fertilizer, more is not necessarily best. Too much can injure plants. I had a garden helper who nearly killed my container plantings by overfertilizing them, so I have been happy with my less-is-more approach ever since. If I remember to fertilize at all, I fertilize little and often. Some of the best fertilizers for container-grown bulbs are liquid fish or seaweed emulsion. These are easy to mix up into a watering can. I suggest using half the recommended strength and observing your plants.

Hardy spring bulbs have all that they need to flourish in their first growing season. They will not need fertilizer. If you want to keep them growing in the same pot for next year, fertilize them while they are in active growth.

Sun and Shade Requirements

Like all plants, bulbs have specific sun or shade requirements. Before you assemble the bulbs into your container, check how much sunlight the display area receives. Sunshine has no impact on growth while the bulbs are beneath the soil. But as soon as they come up, they require a certain number of daylight hours if the bulbs are being saved and grown again next year. The amount of sun or shade that the bulbs receive is irrelevant if you are using bulbs as a short-term display in the garden, discarding them once the show has passed.

A good position for spring-blooming bulb pots is in the shelter of deciduous trees that have not yet leafed out. Pots of summer bulbs are vulnerable to overheating and drying out, especially if they are in full sun. Give them some protection by placing them where they will have afternoon shade. If your plants are not doing well, and you think that they need more or less sunshine, move the pot to a better location.

Considerations for After Flowering

Once your containerized bulb display has finished for the season, you have several options. The first option is to keep them in the same pot and look after them until the following year. Alternatively, you can take them out of their pots, and reuse the container for the next bulb display. If you are using bulbs as annuals, they can be composted. If you want them to bloom next year, they could be planted into the ground or kept in a dry, pest-free place until they are replanted.

If you want to keep bulbs in their original pots during the offseason, you should find an out-of-the-way place to store them. Choose an area with good sunlight, so the foliage can send energy back to the bulbs until they go dormant. Before their next growing season, inspect the bulbs and the roots to check that they are firm and healthy. To do this, turn the pot upside down and knock the soil and bulbs out of the pot. If they look fine, leave them in their clump, knock off some of the outer soil, and refresh the edges with new potting medium. Remove any rotten or moldy-looking bulbs that you find. If the bulbs look unhealthy, replace all the old soil with new to remove pests or disease. If the bulbs have grown well and have filled the whole pot with roots, you can split the clump of bulbs, dividing them into a couple of pots.

Lilies do best without their roots being disturbed. They will be fine for a few years in the

1. Large-leaved tropical bulbs like this *Colocasia* benefit from some shade. **2.** This wooden tub of daffodils, including 'Pueblo', can grow under deciduous trees for a season. Perennial forget-me-nots (*Brunnera*) are growing in the ground. **3.** When grown in pots, Dutch irises like these are generally treated as annuals.

1. Twigs and sticks can be inserted into pots to deter browsing deer from eating your tulips. **2.** Crocuses, like this fall-blooming *Crocus speciosus* 'Albus', are good candidates for pot culture, so that the bulbs can be protected from animals. **3.** I store my pots of hardy bulbs in a sheltered area for the cold months and gradually bring them out in spring.

same pot. Shake off the top of the potting medium each spring, adding fresh soil and a top-dressing of gravel as a mulch.

Reducing Pest Damage in Potted Bulbs

There are various pests, large and small, that want to eat your potted bulbs. The good news is it is easier to protect your bulbs in pots than when they are planted in the ground. There are three main ways that your container-grown bulbs can be damaged: animals can dig bulbs out of the pot, munch the flowers or the roots, or slugs and snails can graze young growth.

You can use a variety of different methods to protect potted bulbs against herbivores. Each garden has a unique set of predators that eat at the bulb buffet that we have laid out for them. You may have insects that eat the roots or leaves, birds that peck at your flowers, or mammals that eat any part of the plant. If deer and rabbits are a problem in your garden, protect your potted bulbs by moving them closer to your house. Reduce browsing by inserting prickly sticks or twigs among your most vulnerable bulbs. Some rodents dig into pots to eat bulbs or hide nuts while the bulbs are dormant.

Your stored winter pots will need protection against digging animals like mice and voles. While pots are in storage, it is best to use both physical barriers and chemical repellents in combination. The only way to totally stop digging damage is to fasten wire mesh over the top of your pots.

Slugs and snails like hiding in the moist places under pots, so check both planted and empty ones. If they are a problem in your garden, use sharp-edged grit on the soil surface. It is rough in texture, and slugs and snails are loath to cross it. Copper tape or a round cross-section of copper pipe can be placed around the plant to act as a slug and snail barrier, particularly around tender bulb shoots.

Caring for Your Containers in Winter

Winter care of potted bulbs depends on the type of bulbs you are growing and your weather. In warm climates, many potted bulbs can be left outside all winter, whereas they will need protection in cold climates.

If you want to save tender bulb that are not hardy outside, bring them inside until next spring. They can be stored in their pots if you have the space. Or, for a more compact storage solution, you can take the bulbs out of their containers, wrap them, and stack them in crates or boxes.

When you are ready to remove bulbs from their pots for winter storage, cut off all top growth. Empty the whole lot into a tub or wheelbarrow. Sort through the bulbs, keeping those that look healthy and discarding rotten or moldy ones. The simplest way to store summer bulbs is to keep them in their original pot surrounded by soil. The bulbs and the potting soil should be kept barely moist. Place them in a cool, but not cold, position, where you can water them if they begin to dry out. In the spring, when days lengthen and the weather begins to warm, water them more frequently to restart active growth. As the plants emerge from the soil, add extra overhead lighting to make sure that you have healthy plants to take outside in spring.

Overwintering Spring-Blooming Bulb Containers

Before spring-blooming hardy bulbs will begin to grow, they need a period of a few months of cold dormancy. The required number of weeks of cold varies by bulb type, but it is usually about 10 to 12 weeks. This chilling period sets the flowering process in motion. When the daylight increases and temperatures warm, your bulbs will grow and flower.

In cold winter climates, potted spring-blooming bulbs freeze solid unless you shelter them from the worst weather. Place your containers under cover, so that the bulbs get the cold that they require while minimizing damage to the pots. Look for a covered place in an unheated, or barely heated, garage, shed, cold frame, or porch with an overhang. The pots can be further protected against cold by wrapping the sides with insulating material, like burlap or hessian tied on with string.

During the winter, check regularly to make sure that the soil is moist, but not wet. If necessary, water the pots. When you see the green bulb noses starting to poke through the soil in early spring, move the pots out to their display positions in the garden.

Growing Indoor Bulbs for Winter Bloom

Even the bleakest of winter days can be brightened by growing flowering bulbs inside the house. While your bulbs outside are dormant, these winter-blooming plants are an excellent way to keep the bulb show going. You may see bulbs for sale that say they are ready for "forcing." This refers to the process of bringing bulbs into early bloom using the warm conditions inside the house.

Narcissus papyraceus—paperwhites

Paperwhites are a type of daffodil that flowers easily inside the house. These are often seen for sale in the autumn, listed for winter blooming. They will take about four to six weeks from planting to bloom. If you want them for a holiday occasion, count back and start them at the right time.

You can plant these up in potting soil, but they grow fine when resting on pebbles in a dish of water. I like to plant them in a decorative bowl by sitting them on colored marbles. When I first started forcing paperwhites, I thought that I could get them to bloom again by planting them outside after blooming. No such luck. They are not hardy in my winters, so the cold killed them. They are hardy to about USDA hardiness zone 8. After they have flowered, put them into the compost heap. If you live in a warm climate, they can be planted outside to bloom next year.

One warning if you have not grown these before: the fragrance of paperwhites is extremely strong. It is a love-it-or-hate-it kind of smell. My eldest daughter loves it, but I find it overpowering especially if you do as I did my first year growing them, planting up dozens and putting them everywhere.

Hippeastrum—amaryllis

Tender amaryllis that are grown inside for winter bloom actually belong to the genus *Hippeastrum*.

Bulbs like this *Scilla forbesii* 'Violet Beauty' are tough and can withstand icy winter weather.

ABOVE: Florist's cyclamen are good for autumn and winter bloom in temperate climates, as seen in this window box. They are also great to grow inside for winter flowers.

OPPOSITE: When it is still cold outside, this non-hardy red amaryllis adds uplifting color inside my dad's back door.

The flowers may be trumpet-shaped or have slender splayed petals. Colors vary, coming in red, white, pink, green, cream, salmon, and various bicolors. They often have a yellow, green, or white center to the flower.

Amaryllis are simple to grow. They are usually grown one bulb per pot. Choose a heavy, deep pot with a hole in the bottom. It should be about a finger width wider than the fattest part of the bulb all around. Put some free-draining potting soil in the base. Then place the bulb on the soil with the pointed end up. It is always worth paying a little bit more to get a nice firm bulb that feels heavy in your hand. Backfill around the bulb, firming the soil as you go. If you like, add some grit or gravel as a top-dressing. Water in and you are done. You can put a saucer under this pot, but do not let water sit there.

Place your amaryllis in a sunny windowsill. When the soil feels dry at the top, give it a thorough watering. To keep the long flowering stem relatively upright, rotate the pot a quarter turn in the same direction every time you water. The flower bud will pop open to reveal a glorious flower. Most people treat these bulbs as annuals. In warm climates you can grow similar-looking amaryllis outside. Try the hardy *Hippeastrum ×johnsonii*, which blooms with red flowers in late winter or early spring, and is hardy in zones 8 or warmer. The similar-looking, summer-blooming, pink-flowered *Amaryllis belladonna* does well in Mediterranean climates.

Cyclamen—cyclamen

There are other bulbs you can purchase already in bloom to grow in the house, such as the tender florist's cyclamen, *Cyclamen persicum*. Grow these red, pink, white, fuchsia, and bicolor plants for their shuttlecock-shaped blooms above their patterned leaves. Like all potted bulbs, they need thorough watering without being wet. Keep them in a bright windowsill. For longer bloom, keep your plants cool. They may go dormant in the summer. If so, stop watering while they have no leaves. Start watering them again in autumn.

Forcing Other Bulbs for Indoor Enjoyment

You may be able to purchase other bulbs, already potted up, that are perfect for bringing into the house for temporary display. The most successful are hyacinths and small daffodils. Keep these potted bulbs in the coolest place possible to extend their bloom. If you want to try forcing spring bulbs yourself for winter bloom, you will need to chill them for about eight weeks in a refrigerator. Store the bulbs away from ripe fruit. Fruit gives off ethylene gas when it is maturing that can interfere with the growth of bulbs. Alternately, you can buy pre-chilled bulbs that will bloom within a few weeks of planting. After blooming, hardy bulbs can be planted outside into the garden.

The Bulb Container Garden | 291

Planting and tending your bulb display can be a real source of joy.

5

PLANTING AND TENDING YOUR BULB GARDEN

A flower-filled bulb garden is an achievable goal, if you learn some basics about planting and caring for bulbs. Bulbs are some of the easiest plants to grow. If you choose healthy bulbs, plant them well, and look after them, they will bloom profusely. Tending to bulbs involves a few basic gardening activities, including watering, fertilizing, mulching, and supporting. Not every bulb needs every type of care. You will get to know each bulb as it grows and figure out what it needs from you to put on a great show.

Buy Healthy Bulbs

Once you have decided what types of bulbs you want to grow, the next step is to find good healthy ones that are ready to do well in your garden. Bulbs are sold in the dormant, or resting, phase of their lifecycle. The bulb is alive but not in active growth. When they arrive at your garden, they should look healthy and ready to plant. Pick them up and do a visual inspection. Check that the bulbs are clean, plump, heavy, firm, and large for their type. Avoid any that are squishy or moldy.

Order bulbs from a reliable supplier, who will send them at the correct time for planting. For autumn-planted spring flowers, order in the summer. Spring-planted bulbs should be ordered in late winter. You will get the best selection if you order early. Look for bulbs that are labeled

ABOVE: No matter what bulbs you grow, examine them carefully to make sure that they are plump and free from mold, such as these daffodils, hyacinths, scillas, and species tulips.

OPPOSITE: If you want a full look, make sure that you order enough bulbs to fill an area, as has been done in this bed of orange Asiatic lilies.

as being grown from cultivated stock. Avoid wild-collected bulbs because their removal and harvest can destroy natural ecosystems. Do not buy bulbs that are listed as an invasive plant in your area.

How Many Bulbs to Buy

The number of bulbs that you need for a certain area depends on how closely together you plant them and how much bed space you are planting. Estimate the number of square feet you will be planting up to give yourself a rough idea. The quantities listed by vendors are usually for an empty bed. Even if I am adding bulbs to a bed that is already planted up with perennials, I get the same number that is suggested. When it comes time to plant them, I plant them closer than recommended. If I run out of bed space, I plant the leftovers in another bed or in a container.

What to Do When Your Bulbs Arrive

If you have the time, bulbs should be planted as soon as the conditions are right for them. Open the boxes or bags and inspect them to see if they are firm and hydrated. Notify your supplier right away if the bulbs are mushy, shriveled, or moldy.

If you cannot plant your bulbs straight away, find a cool, dry, dark area for storage. Do not let your bulbs freeze or get frosted. Look at the bulbs every few days and rotate them to check for deterioration. Tender bulbs need to be kept warmer than hardy bulbs. If they are too wet, they will rot; if they are too dry, they will desiccate.

When to Plant Your Bulbs

There are two main bulb-planting times during the gardening year. Spring-flowering bulbs are planted in autumn, and summer- and fall-blooming bulbs are planted in the spring. There are a few bulbs that should be planted in late summer for autumn bloom.

The best time to plant your spring bulbs is in the autumn when air temperatures cool down, but the soil still retains some summer heat. This often coincides with the time that deciduous trees are changing to their autumnal colors and beginning to fall.

Planting seasons are busy times in the garden. Prioritize planting the bulbs that resent being out of the ground first. Little bulbs like winter aconites tend to wither during transit, so they should be planted as soon as possible. Some larger bulbs like lilies and fritillaries have no outer protective coating, so they will also need to be planted quickly.

During the autumn bulb-planting season, it is best to plant tulips last. All other bulbs need to get into the soil and start growing right away. Tulips do not start growing roots in the autumn, so there is no rush to get them into the ground. It is best to wait until after the soil has cooled down, but before it is frozen. In warm climates, buy pre-cooled tulip bulbs and plant them in winter.

If you hide your bulbs out of sight in the shed or garage and forget about them for a few weeks, you may wonder if it is still worth planting them. As long as you can get a shovel in the ground or have a spare pot, it is worth giving them a try.

Plant your summer-flowering bulbs when you have had a string of nice spring days. The soil should feel warm to the touch. Plant them out at the same time as you are planting out tomatoes, zinnias, and other warm-season vegetables and annuals.

An old way of finding out whether the soil is warm enough for planting is by sitting on the soil, as if you were having a picnic. If you get chilled, the soil has not warmed up enough for planting. If you are an analytical person, treat yourself to a soil thermometer. In climates that have heavy spring rains, wait until the soil is not waterlogged and muddy. The combination of cold and wet can rot bulbs.

A few bulbs need to be ordered and planted at an odd time of year. They need ordering in midsummer for late summer planting. They will bloom a few weeks later in early fall. These are bulbs like colchicums, fall crocuses, sternbergias, lycoris, and cyclamen.

Soil Types and Amendments for Bulbs

The majority of garden bulbs need moist but well-drained soil. If you do not have ideal conditions, there are a few ways to improve your soil. If you are not sure what your soil is like you can dig a hole down to a spade's depth to investigate. Feel the excavated soil particles between your fingers.

If the particles are small and smooth between your fingers, you probably have a clay-based soil. Clay soil holds on to lots of water in many little pores and is too wet for bulbs unless it is improved. To amend clay soil for bulb planting, add lots of large particulate inorganic sand, grit, or gravel, and plenty of organic matter. These soil additions will help increase drainage.

If you rub the soil and discover that it feels like granular, gritty grains similar to table sugar, then your soil is silty, sandy, or rocky depending on the size of the particles. These rock-rich soils are great for growing bulbs from mountains because water drains away quickly, but the soil may still need improving. Sandy soil needs plenty of organic matter added, but it usually does not need extra grit.

Organic matter is the magic ingredient in good, rich garden soil. Many garden bulbs really benefit from it. As organic matter is broken down by the soil invertebrates and micro-organisms, it becomes humus. Humus holds water in the soil and slowly releases it to the bulbs. It is also a natural slow-release fertilizer. Mulching an area with leaf mold or compost is a great way to incorporate extra organic matter into the soil without digging. Bulbs that grow best in highly rocky soils should not be top-dressed with leaf mold.

How to Plant Bulbs

When you are ready to plant your bulbs, get all of your planting supplies together. You will need your chosen bulbs, a planting tool like a shovel, trowel, spade, or bulb planter, some gardening gloves, some grit or horticultural sand, and bonemeal or bulb fertilizer. Bulb planting is easy and pretty much foolproof, even if you have not planted bulbs before.

There are a wide range of specialized bulb-planting tools. I take a simple approach and use my strong and sturdy soil knife. It has a sharp serrated edge that can cut small plant roots in the soil. It is an invaluable tool when planting in an existing flower bed. If I am planting in a new bed or in grass, I use my step-on pointed-tip shovel. There are specialized bulb planters which cut out a plug of soil. If you are using one, have a sturdy stick nearby to push the soil out of the tool. If you have a lot of bulbs to plant, there is a soil auger which attaches to a drill. Take care to use gloves and eye protection when using any tools.

Planting Depth and Spacing for Bulbs

Bulbs have to be planted deep enough in the soil so that they can develop good roots. Each type and size of bulb will need to be planted at the correct depth. Measure the height of the bulb and then dig a hole that is two to three times this measurement. Typically, bulbs are planted deeper in sandy soil than in clay soil, as bulbs are more apt to dry out in sandy soils than in clay ones.

Some bulbs have their own specific planting depths. Large tulips and gladioli need to be planted at a lower depth in order to flower

TOP: Bulbs, like these summer-flowering eucomis, can be planted as soon as the soil has warmed up in late spring.

MIDDLE: Amend the soil with grit when you plant bulbs that need good drainage, like these crocuses.

BOTTOM: A soil knife with depth measurements is a great tool for bulb planting.

well. Some lilies need deep planting to grow stem roots above the bulb. Bulbs like nerines, crinums, and hymenocallis need to be planted at a shallow depth, with their tops out of the soil. Many rhizomes, like bearded irises, need to be planted at the soil surface and only slightly buried. Cyclamen are another example of bulbs that require shallow planting.

When deciding how far apart to plant your bulbs, check the instructions for recommended planting distances. If you choose to plant them closer together, you may have to dig them up and transplant them in a few years. Make sure that your bulbs do not touch each other to reduce the chance of disease spreading. Large bulbs of any type are most likely to need lots of space in between them.

Sometimes you may want your bulb plantings to look like they occurred naturally. This is hard to achieve. If this is your goal, make sure to vary the spacings between bulbs and also between groupings. The old advice to make plantings look natural was to throw the bulbs over your shoulder and plant them where they landed. It can be a guide for you, but often the placement needs finessing. Plan for some concentrated groupings that are surrounded by some smaller satellite trios and singles. Carry on throughout the planting area, varying the number of bulbs in each cluster.

Bulb-Planting Basics

The general rule is that bulbs are planted with the pointed end up. This is easy with true bulbs where you can see an obvious point. Other bulb types may not have an obvious top, so examine the whole bulb carefully. You may see a few shriveled roots clinging to the bulb which may indicate the

BASIC SOIL-IMPROVEMENT MIXTURE

If you think that your planting soil needs to be improved, mix up a wheelbarrow full of fine leaf mold mixed with sand or grit. Use this mixture to top-dress your bulb bed or container by adding a couple of inches on top of your soil. It can be used as a soil amendment anytime you are planting.

bottom. Tuberous begonias and caladiums are planted with the rounded side downward. Some lilies and fritillaries can be planted on an angle to aid with drainage of water out of their scales.

There are bulbs you look at and you are not sure which end should go up. This is often the case with blanda anemone. Just do your best. Usually, the bulbs will correct their orientation as they grow. If you have no clue, try planting them on their side.

To provide extra drainage at the base of the bulbs, plant them on a bed of grit or coarse sand. If you are using bulb fertilizer, then this can go in the base of the hole. If you are planting bulbs that may get eaten, such as tulips or crocuses, add natural animal repellent to the hole too.

When the bulbs are situated, backfill the hole, then gently firm the soil over the top of the planted bulbs with your hands. Bulbs must have close contact with the surrounding soil, with no large air pockets, especially beneath their bases. Small air spaces in the soil are good, so do not use your boots to stamp down the soil. Once they are planted, watering your bulbs helps to settle the soil.

If digging animals are a problem in your garden, you can use some chicken wire or metal mesh to cover the newly turned soil. Hold the wire down with stones or pegs until it is removed before the bulbs shoot up.

Some bulbs benefit from being soaked for a few hours in water before planting to start rehydrating the bulb. You can leave them in water overnight, but not much longer, otherwise they get slimy and begin to rot. Examples of bulbs that may need soaking are winter aconite and blanda anemone.

Coronaria anemone and ranunculus need a little extra care to pre-sprout the bulbs before planting outside. Soak them for three to four hours in room-temperature water. Plant ranunculus bulbs with their cluster of pointed ends facing down into a flat soil-filled tray. Let either of these bulbs grow for a couple of weeks in a cool place until you see roots coming out of the bottom. At this point, you can plant them out into the garden after the last chance of frost. Keep them out of the way of rodents.

TOP: Check the depth requirements before you plant your bulbs. Cyclamen need to be planted at or just below the soil surface.

MIDDLE: For annual displays, bulbs can be planted more closely together than normal as seen here with these tulips in a raised bed.

BOTTOM: Bulbs like these blanda anemones need to be soaked prior to planting.

Keeping Track of Where You Planted Bulbs

Along with keeping track of what bulbs you have ordered, it's a good idea to mark their placement in the bed. This helps you to remember what you planted and stops you from mistakenly digging them up while they are dormant. You can also make a simple sketch of the bed to remind yourself approximately where you planted each bulb. Pebbles, seashells, or decorated rocks are innovative markers that can be used to show the position of bulbs beneath the soil. Knowing the location of buried bulbs prevents you from accidentally spearing them when planting other plants in the same bed.

Insert plant labels deeply into the soil. I suggest always putting them directly behind your planting hole, so you know where the bulbs are when you find the label. I also like to keep a list of bulbs that are in each bed and annotate it with notes about which perform well and those that I must buy more of.

How to Plant Bulbs in Lawns

To plant bulbs in grass, make cuts with a sharp spade into the grass along three sides of an imaginary square. Slice under the turf roots and flap it back as if you were opening a book. Dig a hole in the underlying soil to the required depth. Put any soil amendments, like grit and bonemeal, into the bottom of the hole and plant the bulbs. Lay the soil and the flap of turf back on top and pat it back in place like a golf divot. Planting bulbs in grass may help prevent burrowing animals from digging up bulbs.

Spacing bulbs far apart is a great idea where you want them to perennialize in the ground for years. True bulbs will grow into each other as each individual bulb increases in girth. They may also make daughter bulbs. This is especially true for daffodils. You want them to be able to grow and multiply for years before they need digging up and dividing. Make sure that you leave an unplanted area in your lawn that you will be able to mow as a path or edge. This will give you somewhere to walk and a way to define the planting zone.

There are three basic things to remember not to do when planting bulbs in grass. Do not fertilize the lawn where the bulbs are planted, otherwise the grass will become too strong and may outcompete the bulbs. Do not use herbicide because, depending on what type you use, it may have a detrimental effect on the bulbs. Do not mow for at least six to eight weeks after the bulbs have flowered to allow the bulbs to regenerate for next year. As long as you follow these guidelines, you should succeed.

Watering Your In-Ground Bulbs

After you have planted your bulbs, water the soil thoroughly. The moisture from the soil gets absorbed into the bulbs and initiates growth. Continue watering your bulbs while they are in active growth. Stop watering them once their leaves have fallen off and they go dormant. In some climates that have regular rainfall, there may be no need to water bulbs, even when they are in active growth.

1. I have a collection of historic daffodils that grow together in a bed. To be able to tell which one is which, I have a map of their placement in the bed. To help with identification, they are planted alphabetically from left to right. **2.** Alliums bloom later in the same bed and come up between the clumps of daffodils. **3.** Cluster different types of bulbs together as you plant them in grass to make little pictures, like these 'Elka' daffodils combined with blue scillas.

While bulbs are growing, check whether they need water by looking at the color of the soil and feeling it with your finger. If the soil is dry, water it thoroughly until it is wet several inches down. Deep watering encourages the development of good, long, downward-growing roots. Watering little and often is unhelpful because it brings the roots of the bulbs up toward the surface. Bulbs with extensive roots hold themselves firmly in the ground and are more resilient in times of drought.

It is best to water when the sun is not beating down on your plants. Good times to water are early in the morning or in the evening. Depending on their lifecycles, bulbs have different watering regimes throughout the year. Those that need the same watering schedule and volume should be planted in the same bed.

Summer bulbs grow fast and have large leaves which require significant amounts of water. Cannas, colocasias, alocasias, and dahlias are especially thirsty plants and should be planted within easy reach of a water source. Drought-tolerant bulbs like triteleia, tulbaghias, and some species tulips can thrive with little to no additional water in most climates.

Beware of automatic in-ground irrigation systems in beds where bulbs are planted, as excess water in the soil can harm bulbs. It is best to plant them in places where you can control the amount and frequency of water application. Other wet areas that can cause problems for bulbs include low spots that puddle after rain, such as the bottom of slopes and areas near downspouts.

Summer bulbs, like these large-leaved colocasias and tall hedychiums, can be planted in the same bed. They both require plenty of moisture in the soil as they grow, as shown here at Brent and Becky's display garden, Gloucester, VA.

If you garden where winters are wet, it may be best to dig up summer bulbs and store them inside, even if they could survive the cold in your area. It is the combination of moisture and low temperatures that causes some bulbs to die.

Fertilizing Your Bulbs

Fertilizer can be added to soil to provide nutrients for bulb growth, and is only used when bulb roots, leaves, and flowers are growing. Many soils, especially clay soils, are already fertile. The addition of plenty of organic matter means that you are slowly feeding the soil, which in turn feeds the plant. I am offering this advice about how to fertilize if your plants show signs of decline, poor flower size, yellowing leaves, or small plants. My approach to fertilizing is what I call a "lean and mean" approach. I do not add extra fertilizer except in the hole when planting. The perennial bulbs then have to work hard to get their roots down and seek for nutrients. They develop symbiotic relationships with soil fungi such as mycorrhizae. The resulting top growth is tough and stands upright well. In the following years, I closely monitor my in-ground bulbs to see if they need additional fertilizing.

Relying primarily on the natural breakdown of organic matter to enrich the soil means that it happens slowly but steadily. This mimics what happens in nature, which works well for naturalized bulbs in woodland shade gardens, where snowdrops, wood anemones, and cyclamen grow. Container plantings need more frequent fertilizing because of their limited soil volume. If you are growing bulbs for an annual display, or as spring bedding plants, there is no need to fertilize them because the bulbs will be removed after flowering. Summer- and fall-flowering tender bulbs will need fertilizing because of their rapid growth rate.

If you feel that your plants require fertilizing, there is a range of powdered, granular, and liquid fertilizers. Carefully check the nutrient content before you buy them and, if possible, look for ones that are organic. Not all fertilizers are good for bulbs, especially those with high levels of all nutrients. Too much fertilizer can make the leaves look like they were burned, produce white salty residues, or even kill bulbs. If you suspect that you have over-fertilized your bulbs, add lots of water to flush the fertilizer away and out of the soil. The best practice is to be careful and selective with any chemical additions to your garden and think carefully before you add anything.

On the packet or bottle of fertilizer, you will see three numbers that show you what it contains and at what strength. These numbers each represent the amount of the main elements that are important for plant growth and health. They are listed in this order N:P:K, nitrogen, phosphorus, and potassium. These three numbers are listed by percentages. For example, 10:10:10 would mean that all three are present at 10 percent by volume. There are also micronutrients and minerals listed on the bag.

N stands for nitrogen, a key component of green leafy growth, so add this to your soil for big leafy bulbs like colocasias and alocasias. P is phosphorous, which boosts general plant health and flower production. The last letter, K, is the chemical symbol for potassium, which promotes good root growth. The last two elements are best for promoting good overall growth. These can be used for all spring-flowering bulbs and for summer bulbs that are grown for their flowers.

In a bulb garden, if you need to fertilize at all, use a good organic all-around fertilizer.

Bonemeal is an old-fashioned fertilizer with a ratio of about 3:9:0, which is good when you are planting spring-blooming bulbs. You can also find general fertilizers that are specifically labeled for use with bulbs. Little and often, rather than all at once, is the best way to fertilize.

The fast growth rate of many summer bulbs means that these are the most likely of all bulbs to need fertilizing, especially when they are planted into a container. When fertilizing summer bulbs, diluted liquid fish and kelp fertilizers can be added to your watering can. These are good additions to your watering regime. If you add too much fertilizer at one time, any excess that is not used by the plants gets washed away to the groundwater, especially during a heavy downpour. Follow safety measures when using any fertilizers, taking special care to use gloves and not to breathe in anything.

Mulching

Top the bed with a generous, but not excessive, layer of mulch. Mulches are used in a bulb garden to insulate the soil from extreme temperatures, keep moisture in the soil, and improve soil structure. I suggest adding this little and often rather than a thick layer once a year. Suggested materials for mulching are split into two main groups: inorganic and organic.

Inorganic materials are any type of chipped or rounded stone. If possible, try to source stone from nearby, so that the look of the material is in keeping with your garden. Local stone blends with your environment and fewer shipping miles are needed. Stone and gravel mulch insulate the soil from drying winds, so it is generally moist below the stones, even after weeks of no rain.

1. Tulips that are removed after flowering, like these, do not need added fertilizer in the soil. **2.** Summer bulbs, like these cannas and dahlias, require regular fertilizing. **3.** *Tulipa stapfii*, like many species tulips, benefits from being mulched with gravel. This provides added drainage and some protection against burrowing animals.

Inorganic mulches are especially appropriate for bulbs that come from areas that naturally have dry soil, like rock garden irises, crocuses, species tulips, and triteleias.

Organic-rich materials are things like shredded hardwood, mushroom compost, aged manure, leaf mold, and garden or municipal compost. One of the best mulches for in-ground bulbs is homemade compost or leaf mold, as you know exactly what it contains. Second best is a locally sourced organic product that has only traveled a few miles. Peat is not on my list because harvesting peat from bogs has negative ecological impacts.

Every autumn, bulbs that come from deciduous woodlands naturally receive a layer of leaves that fall on the soil from the trees above. This organic matter decomposes over time and enriches the soil, providing extra nutrients and holding in moisture. Plants that need this moist, well-drained soil include snowdrops, bluebells, and wood anemones.

Thick layers of mulch applied for the winter prevent bulbs from being damaged by cold weather. Mulches are especially helpful where there is no consistent snow cover and temperatures fluctuate widely between day and night. The freeze-thaw cycle is especially harmful to small bulbs that can be pushed out of the ground, where they could be killed.

Weeding

Any plants that you do not want growing in your flower bed alongside bulbs are considered weeds. They use up nutrients and moisture that otherwise could go to your bulbs. Removing weeds early on in their growth is a great way to stop them flowering, setting seeds, and making more weed plants in your bed. Adding a thick layer of mulch will keep weed seeds from germinating.

There are some bulbs that are considered weedy and can become invasive in certain parts of the world. These bulbs set large amounts of viable seed that can spread throughout an area. If they are non-native bulbs, they may replace native plants if they escape from gardens. This is not a problem if they are native where you garden. Check with your local extension service or horticultural experts for the invasive status of bulbs before you plant them.

Allium and ornithogalum are both on invasive lists in some areas of the world. *Allium* 'Hair' was one cultivar that I planted and then have regretted it ever since, due to the aggressive way that it spread. There are parts of the world where you cannot buy alliums for this reason. *Ornithogalum*, especially *O. umbellatum*, has widely escaped from gardens. Its narrow green foliage is grass-like and is easily disguised in lawns, until its white starry flowers emerge in spring.

Another European native bulb that has become a dreadful weed in North America is lesser celandine. It used to be called *Ranunculus ficaria* and is now known as *Ficaria verna*. In early spring, it produces shiny yellow, buttercup-like flowers before it retreats belowground. Corydalis is another bulb that may become a problem in some gardens. The named cultivars are less aggressive than the straight species. *Corydalis incisa* is invasive in large parts of eastern North America.

Crocosmia and agapanthus are examples of bulbs that are difficult to grow in gardens that

In my garden, I use galvanized buckets with holes drilled in the bottom, into which I toss weeds. This bucket is tucked behind a gorgeous 'Black Beauty' lily.

have cold winters. However, in temperate climates, they can take over natural areas and displace the native vegetation as they have in New Zealand, the West Coast of the United States, Ireland, and parts of the United Kingdom. This example reinforces the idea that one gardener's weed is another gardener's treasured plant. Limit your bulb choices to ones that do well in your garden but do not become invasive.

Supporting Top-Heavy Bulbs

Some bulbs are rather top heavy and will need a means of support to stay upright. When done well, the structures that hold up the bulbs should either be a deliberate part of the overall garden picture or be unobtrusive.

Decorative pyramid-shaped supports called obelisks or *tuteurs* can help support tall bulbs. You can make one yourself, using stakes or canes that are inserted into the soil and tied with twine at the top. Place the structure over the plant while it is small, so that the plant can grow up through it. Be careful not to impale the bulb itself. If the plant has already started to grow, then insert the structure into the ground behind the bulb and tie the plant to it.

I use brightly painted wooden poles to support my dahlias. They add a lovely pop of color even before the dahlias begin flowering. You can theme your poles to complement or match the flower colors or use the same color throughout to unify your garden. Use colored twine to tie the dahlias to the stakes to add more color.

Gardeners who like a natural look can stake their bulbs in a way that is an integral part of the garden picture. Wood or twigs can be woven or tied together to create a decorative as well as supportive framework to hold up bulbs. In the flower bed, the color and texture of twigs and small branches blends into the background of the greenery.

Be diligent about staking tall summer bulbs with lots of foliage or heavy flowers, like gladioli and dahlias. They can also be tied to supports or structures like fences. As they grow, tall plants that are not self-supporting need to be tied onto a structure, with twine or garden string. Use a biodegradable string like jute, hemp, or wool, and make a figure-eight pattern with the bulb stem in one loop and the stake in the other. This pattern of string allows some movement of the plant and prevents the string from digging into the stem. If you want to keep the string at a certain height, you can knot it around the stake before tying in the stem. As the plant grows, secure the string by tying with a bow, so you will be able to readjust it later.

Gladioli, especially those with double flowers, need tying into their stakes a couple of times, or need to be grown close to a supportive fence. We like to grow them up through other plants that can help hold the gladioli stems upright. Lilies vary in stem strength with most being self-supporting or supported by surrounding plants. Really tall ones with weak stems benefit from tying in early and often. Dahlias that are grown in borders or for cut flowers will need staking, whereas edging or bedding dahlias do not require staking.

Putting the stake in at planting time is a great idea because you are not at risk of piercing the bulb. Your choices are to put in a stake that is tall enough to carry you to the full height of the summer plant or to put a short stake in as a

1. Bulb supports can be fun, as well as utilitarian. I like to use painted pink poles in my cutting garden to support dahlias, such as this one called 'Ali Oop'. **2.** These unobtrusive natural supports are made of intertwined twigs that will help hold up the leaves and flowers of bulbs, like this crocosmia. **3.** Gladioli can be top heavy when they are in full bloom. Some gardeners grow them against fences for support.

1. Deadheading improves the appearance of summer-blooming bulbs, like this large-flowered 'Lemon Berry' begonia. **2.** I like to leave the seedheads on my snowdrops, shown here among rock garden irises. This allows them to set seed and spread around the garden. **3.** At the end of the season, the foliage of bearded irises needs to be cleared away and should not be composted. This is especially important if you grow a collection of irises.

placeholder for the later full-sized one. As the stem grows, remove the short stake, and replace it, in the same hole, with a taller one. This second method gets around the problem of having a six-inch lily shoot next to an eight-foot pole.

To ensure that the stem does not crack, keep on tying in the plant as it grows. Efficient staking ensures that the stem grows upright and you can see the blooms. By the end of the season, you may have several tiers of twine. Shorter plants can be supported using peony rings, tomato cages, single-stem loops, or semicircular metal supports. If you are growing bulbs for cutting, you can stretch a grid of netting over the bed, letting the plants grow through it.

If you have a windy garden like on a rooftop, balcony, hillside, or by the seaside, supporting tall plants is essential. Bulbs with large leaves like cannas, colocasias, and alocasias that could catch the wind should be sited in a protected spot. Find the direction of the prevailing wind and situate your plants in a sheltered position beside a fence, porch, or deck. Another option is to choose short versions of your favorite bulbs.

Deadheading

Deadheading is the process of removing the old flowers of your bulbs after they have finished blooming. This prevents the plants from using energy to set seed. Instead, the energy will be sent down to the bulb to fuel next year's flowering.

Daffodil seedheads are easy to snap off between your thumb and forefinger once the flowers become papery in texture and have turned brown. If tulip bulbs are going to be composted after bloom, there is no need to remove seedheads. Little bulbs are too small to deadhead.

Dahlias need regular deadheading to encourage the plants to continue flower production. Cannas produce several flowers to one stem, so carefully take off the one flower that has finished and leave the rest to bloom. Begonias, especially the large-flowered ones, will need careful removal of the old blooms. If you are not saving their seeds, lilies can be deadheaded.

There are times when you want to leave the seedheads on your bulbs. If you do not deadhead, the seeds can mature and be dispersed around your garden. As these seeds grow in the next few years, they can develop into full-grown bulbs. This is most often seen in bulb lawns with little tommie crocuses that spread well by seed to create large swaths. Most people do not bother to deadhead small bulbs. I never deadhead snowdrops so that they seed in profusely. I am delighted by the resulting random placement and increased number of flowers.

Neatening Up Bulb Foliage

Once spring-flowering bulbs have finished blooming, their foliage will continue to absorb sunlight for a few more weeks. During this time, it is important to let them finish their growing cycle, so that they will come back and flower next year. After the bulb leaves wither and turn brown, gently tug at them to see if they come away from the bulb easily. If so, gently clear them away and compost them.

Do not "tidy up" your hardy bulb foliage by cutting it off, tying it up into knots, or braiding it. Instead, guide the leaves away from neighboring plants, using short sticks or bamboo canes, until the foliage can be gently removed. If you plant your spring bulbs among perennials that grow strongly in late spring, the withering leaves will be slightly hidden but can still get the necessary sunlight. Suggested perennial companions are catmint, border phlox, asters, daylilies, peonies, and salvias.

The foliage of bulbs that flower separately from when they grow leaves (like lycoris, stern-bergias, and colchichum), should be left in place until it falls away. Summer and fall bulbs also need strong leaf growth. They need to have their leaves tidied up if they turn brown or suffer other damage. Do not compost any foliage that may have pest or disease issues. Bearded iris, dahlia, and lily foliage should not be composted.

Special Bulb Maintenance for Summer Bulbs

Some summer bulbs require special treatment in the garden. It doesn't mean that these bulbs are harder to grow than others, but if you know what they need, they will perform well for you in the garden.

Dahlias

To grow dahlias well, there are a few specific things to know about planting and maintaining them. When you plant your dahlias, make sure that the soil is warm and not too wet. Dig a hole that is deep enough, so the bottom of the clump is six inches down. Lightly cover the dahlia with soil and backfill the hole as the shoot grows. This

helps to stabilize dahlias, especially if they are tall. Do not start watering until the dahlia sends its first shoots above the initial soil covering. If you garden in a cool summer climate, it is often best to start the dahlias into growth inside. This way, you get a faster start when the weather warms up.

When the plant has reached about 18 inches in height, take out the growing tips of the primary dahlia shoot. This initiates the growth of side buds and produces a full-looking plant. Dahlias that are three feet or more in height should be staked when they are planted. All dahlia branches need to be regularly tied to their support. Water and fertilize dahlias during the growing season. They require lots of water to grow to their full potential. Pick dahlia flowers regularly and deadhead old ones promptly. Flower buds will continue to be formed until short autumn days slow down growth. If dahlias are not hardy in your area, they will need to be dug up for the winter. Leave them in your garden until there is a frost and then dig them up a week later. Turn them upside down to drain water out of their hollow stems. Then bring them into their winter frost-free storage.

Cannas, Alocasias, Colocasias, and Caladiums

These tender bulbs can be started into growth inside in spring, while the outdoor conditions are still cold. This gives them an early start on growth, so you will have maximum summer show. They must only be planted out when the soil is warm and there is no possibility of frost. Place them in

Dahlias are glorious flower producers but need some specialized plant maintenance. Here, 'Ryecroft Jill' dahlia is grown in a display garden with other types behind.

a location with bright light, but that is not in hot, late-afternoon sun. Caladiums need more shade than the other plants listed. However, there is a range of sun-loving caladiums that you could seek out. Watering and fertilizing are essential to good growth. During the summer, take off old leaves or flowers. Dig these bulbs up before the extreme winter cold begins. Keeping them from year to year will result in larger plants next year. They can be divided into separate clumps to make more plants.

Gladioli

Gladiolus hybrids are one of the showiest bulbs to add to your summer garden. Plant them about six inches deep in rich soil with added compost. Deep planting helps the stalks to self-support, but tall ones might need additional staking. For a sequence of blooms, plant a few bulbs every couple of weeks from late spring to midsummer. Water them regularly during the growing season. When they are about to flower, add liquid fertilizer with high potassium. Take the spent flowers off the stem as they continue to bloom from base to top. If you are cutting the flowers to use for arrangements, and you want to use the same bulbs next year, make sure you leave on some foliage. If the bulbs are hardy in your area, mulch the ground above them with organic material for the winter. In cold climates, they will need to be lifted and stored in a frost-free place.

Lilies

Lilies require really good drainage around the bulbs. If you cannot plant the bulbs as soon as you receive them, keep them in a cool dark place for a short time. They need to be planted at least four to six inches deep. Sit the bulb on grit and backfill with grit-enriched soil. Once the young shoots emerge, make sure to protect them from frost and slugs. Lilies need water until they bloom. After that, many are drought tolerant. If your soil is poor, use a low-nitrogen fertilizer. If the lilies are planted in full sun, they are unlikely to need staking. If you are growing them for arrangements or want to deadhead them, leave as much of the stem and foliage on the plants as possible.

Bulb Propagation: Division and Seed Growing

After a few years, bulbs growing in the ground can become overcrowded. This especially happens with true bulbs that are hardy in your climate. Dividing a clump of existing bulbs is an easy way to rejuvenate an old group or to spread your bulbs to other areas of the garden. Each new division will have better access to water, nutrients, and sunshine.

The best time to divide hardy bulbs is after they flower, when the foliage is dying down. Lever the whole clump up out of the ground by digging around the perimeter in a wide circle. A garden fork or spade is a useful tool for this. Remove the clump of bulbs from the ground and examine it. Discard any bulbs that look rotten or damaged. Gently pull the clump apart into smaller segments. I have found that bulbs re-establish faster if I leave a few bulbs together and replant them in groupings, rather than separating each single bulb. I add some amended soil into the bottom of the hole and firm down the soil surface. Replant some bulbs back in the

1. Most large-leaved summer-flowering bulbs, like this 'Toucan Scarlet' canna, require plenty of water and fertilizer while they are in growth. **2.** Hybrid gladioli, like this 'Wine and Roses', are tall summer bulbs that need organic-rich soil, plenty of water, and good drainage. **3.** Lilies are great bulbs for the summer garden. If you choose short varieties and plant them in full sun, they should not need supports. Here they are grown between alliums and salvias. **4.** Leucojum bulbs are easy to dig and divide, but they dry out easily. Replant them as soon as possible.

original hole with a little bonemeal sprinkled in the bottom. The other clumps can be moved to different places or given away. Bulbs that can be divided this way include daffodils, lilies, lycoris, zephyranthes, hymenocallis, and tuberoses.

There are some hard-to-establish bulbs that tend to fail when you buy them as dry bulbs. Getting a clump from a friend, straight out of the ground, works better. Wrap them in damp newspaper for the ride home. Examples of bulbs that easily dry out include leucojums, snowdrops, and winter aconites. Snowdrops are often sold or exchanged "in the green," which means directly after flowering, with the foliage still on the plant. The advantage to this method is the other snowdrops in the garden still have their leaves on, so you can see where there are gaps that need filling.

Another easy way of propagating some lilies, such as tiger lilies, is by taking the little bulbils that are tucked into the leaf axils and planting them. For the first few years they only grow leaves, stems, and a gradually enlarging bulb. Once the bulb is large enough, a flower stalk is formed. You can harvest mature bulbils and grow them in pots, or in an out-of-the-way place, to increase your stock of these lovely flowers or to swap with friends.

Other lilies are easy to grow from seeds. If you are new to seed-sowing, try starting the easy-to-grow Formosa lilies. You can plant them directly into the garden or in a pot of gritty seed-starting mix. The green leaves grow the first year, and one or two years later, the flower. Some alliums are also easy to grow from seed. Dahlias are available to buy and grow from seed.

Gladioli produce small offsets called cormlets or cormels. These can be planted into a pot until they grow to flowering size, when they can be planted out into your garden. If they are not hardy for you, they will need to be protected from frost.

Troubleshooting Bulbs That Don't Bloom

When you plant new flower bulbs, they will usually bloom the first year. The purchased bulbs come from a nursery or a field where they were growing in ideal conditions. If they do not bloom in subsequent years, one of the following factors may be in play. The major causes of bulb failure are congestion, insufficient sunlight, problems with water, or pests.

ABOVE: The black bulbils in the leaf axils of this tiger lily can be detached when they are mature and planted to grow more bulbs.

OPPOSITE: I have been growing this group of 'Early Bride' daffodils in the same spot for a few years. The clump will need to be divided when the number of flowers starts to decline.

Congestion

When you planted your bulbs, each one had plenty of space. Over time, as they grew and multiplied, the bulbs become one congested clump. These packed-in bulbs can no longer get enough water, nutrients, or sunlight to grow and flower. If this happens to a patch of your bulbs, the solution is to dig up the group and divide them into several smaller parts before replanting them farther apart.

Not Enough Sunlight

The next reason your bulbs may have stopped flowering is, during the preceding year, they did not get enough sunlight energy to power flower production. Surrounding shrubs might have grown up and now shade the planting bed. This is primarily a problem of long-lived bulbs like daffodils and bearded irises. Look at the area where you are growing these bulbs. See if there is a way to give them additional sunlight. Trim some of the surrounding plants or move the bulbs to an area with more sun. It will take a season of good growth before these moved bulbs will flower again.

Problems with Water

Often when a bulb does not come back it is in overly wet soil and has rotted. This may be due to excess irrigation, planting in a wet spot, or failing to amend the soil. The opposite water extreme, that of having no water, can prevent bulbs from

Some species tulips, like this Sprenger's tulip, can be long-lived in the garden and may seed in. These tulips need to be planted deeply. Once planted, they are difficult to move.

making flowers. Bulbs need to be moist, but not wet, during their resting phase. Then more water is needed for the bulbs to grow and flower. For summer bulbs that are stored inside, there may have been storage problems over the winter. Keep them hydrated, but not so much that they rot.

Issues with Fertilizing

If there are inadequate nutrients in the soil or if there is too much nitrogen, summer-flowering bulbs may fail to flower. When nitrogen is added in large amounts, it encourages rapid leafy growth at the expense of blooms. Check your fertilizing regime to make sure you choose the correct one for boosting blooms.

Bulbs That Are Not Perennial

There are some bulbs, such as large tulips, that seem like they should come back in your garden, but actually perform like annuals. They bloom wonderfully in the first year after you plant them. In following years, their bloom might be sporadic. If you would like to grow tulips that flower year after year, choose those that are labeled as perennials, such as species tulips and the Darwin hybrids.

Another disappointing category of bulbs are those you believe to be hardy, but do not emerge after winter. While the hardiness zones are a guide for bulb choice, they are about averages. Your garden may have experienced atypical temperatures, or those special killers of bulbs: freezing and wet. If you live in very hot climates, heat can kill bulbs too.

Damage to the Bulb

Sometimes, the most obvious reason that your bulbs are not flowering is they have been eaten or otherwise damaged. Deer and rabbits are insatiable and voles and burrowing animals are very hungry. If it is only the flower itself that has been eaten off, the bulbs may flower next year. If the bulbs themselves have been consumed, you will need to replace them and replant in a protected area. Runoff from neighboring lawns that contains excess fertilizer or herbicides is another way that your bulbs can be damaged.

Winter Storage of Tender Bulbs

In gardens that experience warm winters, most bulbs can stay outside year-round. If you garden where winters are cold, all tender bulbs that you want to save will need to be dug up and brought inside until the next growing season. You may want to experiment with borderline-hardy bulbs by adding a thick layer of mulch on the soil surface. To keep it in place, encircle the mulch with wire mesh.

Bulbs that are not hardy in your climate will either need to be composted at the end of the growing season, or dug up and saved in a frost-free place over winter. The storage area should be above freezing, preferably in the range of 40 to 50 degrees F, as these bulbs come from the tropics or subtropics. Store the bulbs where you can easily get to them and water them as necessary. Suggested places for winter storage are a garage, basement, slightly heated shed, or a sun porch.

TOP: Borderline-hardy bulbs can be protected by mulching them. Here, a ring of chicken wire keeps the mulch in place above the crown of a dahlia, insulating it against a late spring hailstorm.

BOTTOM: Bulbs like these acidanthera gladioli can be dug up in the autumn if they are not hardy. Store them in a frost-free place over winter. These bulbs have little cormlets attached to the bases, which can be planted to make new bulbs in a few years.

If you want to overwinter your tender bulbs in cold climates, dig up your bulbous plants in autumn as the growing season draws to a close. There are two categories of tender bulbs. The ones that are truly tropical will start to appear distressed as night temperatures dip. As soon as you see signs of decline, move them inside. The other group can take some cold and is brought inside after the first frost. The foliage may have been nipped by frost or it could be turning yellow.

To remove them from the ground in autumn, take a large shovel and start to dig far away from the base. Make a circle around the clump, levering the group gradually out of the ground so you do not damage the roots. Cut off the tops, leaving a foot of stem as a handle. Let the roots dry out for a few days in a frost-free area. Turn them on their sides or even upside down, allowing any water accumulated in the neck or leaves to drain away. I use a crate with holes in it to balance the bulbs. A cardboard box works just as well.

Wrap the bulbs in newspaper or use a box or crate filled with shredded paper. Store them in your cool, but not cold area, checking the roots regularly for drying out or rot. If they are dry, dampen the newspaper. If they are too moist, open up the paper to let them get air. You will learn how often you need to check them. I have a routine of checking all my stored bulbs on Saturday mornings, when I water my houseplants. Make sure that rodents are excluded from the storage area.

Pests, Diseases, and Herbivores

Pests, diseases, and herbivores can all impact healthy growth of bulbs. In a home garden, pests and diseases may not play a huge part in your day-to-day bulb care. Herbivores, however, can be a big problem, but you can take concrete steps to limit damage.

For the best chance of success, buy healthy bulbs that suit the growing conditions in your garden. Be observant while your bulbs are growing. If you notice anything out of the ordinary, investigate further. If you can tolerate a small amount of pest damage to your bulbs, then little

action will be needed. Remember all of the pest and disease actions that are happening in your garden are part of a bigger natural ecosystem, with many normal lifecycles taking place there. You might want to do something about pests when this system of checks and balances gets out of control, but sometimes it's good to let go of perfection. The best way to grow healthy plants is to choose the right bulbs, plant them correctly, and look after them well. Plants that are growing vigorously are likely to resist diseases.

Another strategy is to grow a diverse range of bulb types. Rotate annual bulb plantings into different areas or beds, in the same way as you might do with vegetables. Bulb pests and diseases are often specific to a particular genus or family of plants. If the whole bulb or plant is eaten, grow it in another place where it is protected, or choose a different type of bulb in its place.

Bulb Diseases

There are some bulb diseases caused by fungi, viruses, or bacteria. If you buy disease-free bulbs and grow them in well-drained soil, you are unlikely to have major problems. One disease you may encounter is powdery mildew. This disease makes the leaves of some summer bulbs, like dahlias, look like they have been dusted with flour. Make sure the roots are well hydrated. Water the soil rather than the leaves in the morning, so that the plants dry off during the course of the day. There are other diseases which can affect particular bulbs. Most of these are encountered when you grow a lot of one type of bulb in close proximity to each other. If you suspect that a plant is diseased, remove it and dispose of it carefully. Do not put it into the compost. If in doubt about your disease problems, contact your local garden center or other horticultural advice service.

Insects and Other Arthropods

Your garden is a busy place containing a multitude of insects, many of which are either beneficial pollinators or are just passing by. If you notice damage, don't assume that an insect has caused the problem. Some blemishes may be due to cultural or environmental issues, such as a late frost that damaged leaves or a lack of moisture in the soil at a crucial time. Anytime a plant is stressed, it is more susceptible to damage.

Signs of insect damage include stunted growth, sticky residues, or leaves that are eaten, yellowed, or mottled. Turn the leaves over and see if there are obvious insects. Try to identify them to make sure they are not beneficial insects coming in to eat the problem insect. I never use pesticides in my garden because they are bad for the beneficial insects, and I tackle small outbreaks fast. If I identify a problematic insect, my first line of defense is to spray the insects off with a sharp stream of water from my hose. I also pick off or squish ones that I know are causing damage.

Aphids are common garden insects that suck sap and exude a sticky substance, called honeydew. The honeydew sticks to the leaves where the insects are feeding, and a black mold grows on the honeydew. This black mold might be the first sign of the presence of aphids. Other insects that can cause problems for your bulb plants include leaf hoppers, which can spread viruses and other diseases, and earwigs, which can nip off leaves or make holes in flowers with their sharp pincers. The old way to catch earwigs was to make a bundle of straw inside a small pot, balancing it upside down on top of a plant stake, near where you have noticed issues. Once earwigs make themselves at home in there, you can remove the pot to a faraway place.

If you grow bearded irises, the pest you need to watch for is the iris borer. This is a larval stage of a moth that tunnels into rhizomes and eats into them, creating holes. The damage can be an entry point for other problems. Another telltale sign of these moths are brown streaks in the young leaves that are in active growth. To reduce the incidence of iris borers in your garden, use clean culture methods like clearing the foliage away at the end of the season and disposing of it, not composting it. Dig up your irises every few years, in the late summer, to inspect the rhizomes for damage. If possible, replant them elsewhere. Ensure that the area gets plenty of sun; plant the rhizomes at the soil surface; and add more gravel to the planting bed.

One beetle that causes havoc in my garden is the bright red lily beetle. It eats the leaves of any bulb in the lily family, including true lilies and fritillaries. The telltale signs are jagged notches taken out of leaves and blobs of black slime under the leaves that cover the juvenile insects. Their eggs are shiny, tiny, and laid on the undersides of leaves. If the population explodes, the adult beetles defoliate entire plants. A few lilies like 'Black Beauty' and 'Scheherazade' show some measure of resistance. To reduce the population, pick off the adults and crush them. They have a black underbelly, which allows them to hide on dark soil. Catch them before they fall or place white paper beneath the plants to make the beetles obvious. Remove and dispose of eggs and the black slime–covered larvae. Do this every day while lily beetles are active.

Another plant- and flower-eating beetle that causes problems in North American gardens is the Japanese beetle. Canna is one of their favored

1. Depending on their growing conditions, dahlias can develop powdery mildew, as seen on the spent flower of this single 'Delta Red' dahlia. **2.** The majority of insects that visit your garden are beneficial, like these bees on drumstick alliums. **3.** If you grow lilies or fritillaries, look out for red lily beetles. Dispose of them by picking them off the plants.

plants. Gladioli and calla lilies can be affected too. The beetles chew into leaves and flowers, making extensive holes. I catch the beetles or knock them off the plants into a bucket of soapy water. I try to do this daily, in the early morning before the beetles become active.

Spider mites are small spider relatives that can damage some bulbs by sucking on their leaves. Signs to look for are mottled or discolored foliage. The mites are usually found on the bottom of the leaves and can be sprayed off with a strong stream of water from a hose. The main thing to do is to improve the cultural conditions of the plant. Make sure that the plant is watered regularly and deeply. Do not overfertilize.

You may come across other things that you think are pests of bulbs in your garden. First confirm that they are indeed pests. Once you know what they are, you can figure out how to physically remove them from your plants, or how to make the growing conditions better.

Slugs and Snails

Slugs and snails are both types of mollusks. They are part of gardening, especially in areas with high annual rainfall. They tear characteristic ragged-edged holes into leaves with their rasping mouthparts. If you have lots of slugs and snails, the only sustainable way to garden is by planting bulbs that they do not like to eat.

Slugs and snails like hiding in moist places under pots, so check under both planted and empty ones to find them. They can be squished, put into the compost heap, or on the bird-feeding table. If you have a plethora of frogs, toads, and mollusk-eating birds, they will consume many slugs and snails, and generally maintain a good

TOP: Telltale signs of slug damage include holes taken out of flowers, like this snowdrop.

BOTTOM: Encourage frogs and toads to live in your garden. They can keep your slug numbers in check.

population balance within the garden. This natural pest-control method involves accepting some damage, while proactively protecting young plants until they can fend for themselves. Sharp-edged grit on the soil surface can reduce harm to your plants, since it is rough in texture and slugs and snails are loath to cross it.

Handpicking and destroying slugs and snails can reduce the overall populations in your garden. Use a light to find them at dusk and night, since this is when they are active. You can create traps to lure the slugs and snails. Use upside-down eaten citrus halves or shallow cans filled with old beer and set them low in the soil. Empty the traps in the morning.

Be watchful when young tender bulb shoots emerge from the ground. Protect your plants at this stage as mollusks are particularly drawn to munch on this fresh growth. Try physical slug barriers including sheep's wool and copper. Sink a copper section cut from a pipe into the ground around particularly attractive plants like emerging dahlia shoots. Copper tape or mesh around a pot and saucer is another option to try.

Protecting Bulbs from Herbivores

If you garden alongside digging, burrowing, and chomping animals, it sometimes seems like a battle of wills between the gardener and the herbivores as to who will get to enjoy the bulb flowers first. Find out what animal or animals are eating your bulbs. Once you have identified your foes, it is easier to work out how to foil them. Common culprits are deer, rabbits, mice, voles, moles, and other plant-eating varmints. No one solution suits every garden. Use trial and error to see what works for you.

The most reliable way of protecting bulbs from being eaten is to have a physical barrier such as a sturdy fence between the animals and your plants. Use a narrow-gauge wire mesh that is attached to the fence several feet above the soil line and dug into the ground at least a foot. It will not stop all burrowing animals, but it is a good start. Make sure there are no gaps because a determined animal will find the weakest link.

Woven wire mesh can be a useful physical barrier in other ways. Bury the bulbs in a protective wire basket. Or cover the surface of a freshly planted flower bed with chicken wire, held down at the edges with rocks or logs. The gauge of the metal mesh should be small enough to prevent the burrowing animals from gaining access to the bulbs, but large enough to let the bulb shoots emerge through the holes. As the bulbs pop up you can remove the wire on the soil surface.

We use a lot of chipped gravel and grit around our bulb plantings. The sharp edges of the gravel may dissuade some animals from digging, and at the same time, improve drainage. Bury the bulbs deeper than usual in a pocket of grit or gravel. Other gardeners use crushed eggshells as a scratchy physical barrier.

To deter animals from eating your bulbs, you can sprinkle or spray chemicals that smell or taste bad to them. Common constituents of these preventative mixtures include garlic, herbs, eggs, chili peppers, and herbal extracts. Some of them smell so obnoxious that they are a terrible choice if you are sensitive to smells or if you are growing bulbs for fragrance. If you do use them, take necessary precautions. Keep mixtures away from children and pets, wear gloves, and use eye protection. Repellents have a limited lifespan and work until they are washed off by rain or irrigation.

Choose Bulbs That Are Less Likely to Be Eaten

One way to reduce animal problems is to choose bulbs that are unlikely to be eaten. Two of the largest animal problems are rabbits and deer. Both of these types of animals are voracious. It can be difficult to say exactly what they will eat or not eat. Young animals who have not yet learned what to avoid will try to eat anything.

Crocuses, tulips, and lilies are highly attractive to mice, rodents, and other animals. Some gardeners bypass these bulbs altogether. Other bulbs that may be eaten include liatris, dahlias, and lycoris.

If you are looking for bulbs to grow that are considered resistant to deer and rabbits, and also burrowing animals, try growing snowdrops, daffodils, hyacinths, scillas, fritillaries, and alliums for spring bloom. For later in the year, choose colocasias, calla lilies, tulbaghias, sternbergias, and colchicums. This list is based on my own experiences and from talking to other gardeners. These bulbs contain substances, like toxic alkaloids or oxalates, that herbivores avoid. You can also try interspersing tulips with daffodils to reduce damage. Or you could grow vulnerable bulbs among strongly scented herbs, like sage or rosemary, as a natural repellent.

1. Grow colchicums and other bulbs that are unlikely to be eaten if you garden alongside a variety of different herbivores. **2.** The easiest way to protect vulnerable bulbs from being eaten is to fence in an area against rabbits and other herbivores. **3.** Hide bulbs that might be eaten by surrounding them with plants that have odors, like most herbs. Here, regal lilies are growing with feverfew (*Tanacetum parthenium*) and mountain mint (*Pycnanthemum*).

EPILOGUE

The End Is the Beginning

As this bulb book draws to a close, my wish for you is to grow more bulbs. They are such a rewarding group. There is a bulb or two that would suit each garden and every gardener. If you add a few dozen here and a handful there, you will have a garden full of bulbs before you know it. Continue to choose ones that do well for you and that you love. Include more of the tried-and-true bulbs that are the mainstays of your garden. Then, every year, try a couple of new ones. This approach gives your garden continuity combined with freshness.

This procession of dazzling bulbs has come to an end. If you are reading this book backward, like I often do, welcome to this bulb book. If you started at the other end, the front, I hope that your gardens and dreams are, or will be, filled with a bounty of beautiful bulbs, both old and new.

At the end of the day, I even enjoy the fallen petals of this lovely 'Apricot Parrot' tulip.

SUGGESTED READING

Daffodils for North American Gardens by Becky and Brent Heath. Houston, TX: Bright Sky Press, 2001.

The Gardener's Guide to Growing Dahlias by Gareth Rowlands. Portland, OR: Timber Press, 1999.

The Gardener's Guide to Growing Fritillaries by Michael Jefferson-Brown and Kevin Pratt. Portland, OR: Timber Press, 1997.

The Gardener's Guide to Growing Irises by Geoff Stebbings. Portland, OR: Timber Press, 1997.

The Gardener's Guide to Growing Lilies by Michael Jefferson-Brown and Harris Howland. Portland, OR: Timber Press, 2002.

Glorious Shade: Dazzling Plants, Design Ideas, and Proven Techniques for Your Shady Garden by Jenny Rose Carey. Portland, OR: Timber Press, 2017.

The Plant Lover's Guide to Dahlias by Andy Vernon. Portland, OR: Timber Press, 2014.

The Plant Lover's Guide to Tulips by Richard Wilford. Portland, OR: Timber Press, 2015.

Snowdrops: A Monograph of Cultivated Galanthus by Matt Bishop, Aaron Davis, and John Grimshaw. Cheltenham, UK: Griffin Press, 2006.

The Tulip: The Story of a Flower That Has Made Men Mad by Anna Pavord. London: Bloomsbury, 1999.

Tulips for North American Gardens by Becky and Brent Heath. Houston, TX: Bright Sky Press, 2001.

The Ultimate Flower Gardener's Guide: How to Combine Shape, Color, and Texture to Create the Garden of Your Dreams by Jenny Rose Carey. Portland, OR: Timber Press, 2022.

CONVERSION TABLES

inches	cm
1	2.5
2	5.1
3	7.6
4	10
5	13
6	15
7	18
8	20
9	23
10	25
20	51
30	76
40	100
50	130

feet	m
1	0.3
2	0.6
3	0.9
4	1.2
5	1.5
6	1.8
7	2.1
8	2.4
9	2.7
10	3
20	6

Temperatures

$^\circ C = \frac{5}{9} \times (^\circ F - 32)$

$^\circ F = (\frac{9}{5} \times ^\circ C) + 32$

ACKNOWLEDGMENTS

The greatest of thanks always goes to my extended family for their support during the book-writing and photography process. To the following people I express my deep gratitude: to my very supportive and wonderful husband, Gus, who kept me supplied with daily coffee and chocolate treats; to my three lovely daughters and their husbands, Meade and David, Janet and Anastas, and Emily and Jake, for lots of garden visits paired with advice and encouragement; to my two wonderful grandsons, Oli and Arthur, for extra hugs when needed; and to my family in England, for more garden trips and suggestions from that side of the Atlantic.

To Helene Fantini-Cooper, thank you for your wonderful editorial help and for keeping me organized for garden lectures and visits. Big thanks go to Joe Giampa Jr. for keeping Northview functioning and for adding so much artistic flair to the garden. The garden would not be the same without your carpentry and creative skills. To Hanna von Schlegell, who has played many roles to do with bulbs both in the garden and for this book. She planted thousands of bulbs each year, when the book was a dream. Specifically for this book, she added many more bulbs, both in the ground and in containers, and then staged them for their photography shoots. Hanna has been my right-hand person both in the garden and in editing. She amazingly keeps bulb names straight even when the squirrels steal the plant labels.

I would like to thank Timber Press for their support of this bulb book. Special thanks go to my editor Makenna Goodman, photography editor Sarah Milhollin, and copy editor Mollie Firestone.

Thank you to the following gardens and gardeners. I really appreciate your advice and for allowing me access for photography.

Andalusia Historic House, Gardens, and Arboretum, Andalusia, PA; Glenn Ashton, Meadowbrook, PA; Brent and Becky's Bulbs, Gloucester, VA; Paul and Brady Brown, Three Tuns, PA; Chanticleer Public Gardens, Wayne, PA; Janet Chrzan, Media, PA; Colonial Williamsburg, Williamsburg, VA; Dahlia Dell, Dahlia Society of CA, Golden Gate Park, San Francisco, CA; The Delaware Valley Daffodil Society, PA; Betty Dennis, Broadway, Worcestershire, UK; John and Alex Dodge, Broadway, Worcestershire, UK; Fuller Gardens, North Hampton, NH; Great Dixter, Northiam, East Sussex, UK, Pamela and Duane Hubbard, Effort, PA; Inland Empire Dahlia Society, Spokane, WA; Jenkins Arboretum, Devon, PA; Lady Margaret Hall, Oxford, UK; Elizabeth Lawrence House and Garden, Charlotte, NC; Longwood Gardens, Kennett Square, PA; Manito Park, Spokane, WA; Mt. Cuba Center, Hockessin, DE; Carla Zambelli Mudry, Malvern, PA; Northview Gardens, Ambler, PA; Old House Garden Bulbs, Ann Arbor, MI; Oxford Botanic Garden, Oxford, UK; Presby Iris Gardens, Montclair, NJ; San Francisco Botanical Garden, San Francisco, CA; Waterperry Gardens, Oxfordshire, UK; Wave Hill, Bronx, NY, Winterthur Museum and Garden, Winterthur, DE

INDEX

JENNY ROSE CAREY is a renowned gardener, educator, historian, author, and the former senior director at the Pennsylvania Horticultural Society's Meadowbrook Farm in Jenkintown, PA. She previously worked at Temple University for more than a decade, first as an adjunct professor in the Department of Landscape Architecture and Horticulture, and then as director of the Ambler Arboretum of Temple University. Jenny Rose has been lecturing nationally and internationally for many years. She is an avid hands-on gardener who has gardened in both England and the United States. Her Victorian property, Northview, contains diverse garden spaces, including a cutting garden, an herb garden, a dry garden, and various mixed flower beds. Jenny Rose and her gardens have been featured on the PBS series *The Victory Garden*, in *The Wall Street Journal*, *The Washington Post*, *The Philadelphia Inquirer*, *Grow* magazine, and *The Pennsylvania Gardener*. She is the author of *The Ultimate Flower Gardener's Guide* and *Glorious Shade*, both published by Timber Press. Find out more at her website: jennyrosecarey.com and Instagram @northview-garden and @jennyrosecarey.

Timber Press
Workman Publishing
Hachette Book Group, Inc.
1290 Avenue of the Americas
New York, New York 10104
timberpress.com

Timber Press is an imprint of Workman Publishing, a division of
Hachette Book Group, Inc. The Timber Press name and logo are
registered trademarks of Hachette Book Group, Inc.

Printed in Hung Yen, Vietnam,(APO) on responsibly sourced paper
Text and cover design by Vincent James

The publisher is not responsible for websites (or their content)
that are not owned by the publisher.
The Hachette Speakers Bureau provides a wide range of authors
for speaking events. To find out more, go to hachettespeakersbu-
reau.com or email hachettespeakers@hbgusa.com.

ISBN 978-1-64326-324-3
A catalog record for this book is available from
the Library of Congress.